Guide and Reference to the
Snakes of Eastern and Central North America
(North of Mexico)

University Press of Florida

Florida A&M University, Tallahassee
Florida Atlantic University, Boca Raton
Florida Gulf Coast University, Ft. Myers
Florida International University, Miami
Florida State University, Tallahassee
University of Central Florida, Orlando
University of Florida, Gainesville
University of North Florida, Jacksonville
University of South Florida, Tampa
University of West Florida, Pensacola

Guide and Reference to the

University Press of Florida
Gainesville
Tallahassee
Tampa
Boca Raton
Pensacola
Orlando
Miami
Jacksonville
Ft. Myers

Snakes

of Eastern and Central North America (North of Mexico)

R. D. Bartlett and Patricia P. Bartlett

Library of Congress Cataloging-in-Publication Data
Bartlett, Richard D., 1938–
Guide and reference to the snakes of eastern and central North America
(North of Mexico) / R. D. Bartlett and Patricia P. Bartlett.
p. cm.
Includes bibliographical references (p.).
ISBN 0-8130-2935-X (alk. paper)
1. Snakes—North America—Identification. I. Bartlett, Patricia Pope, 1949–
II. Title.
QL666.O6B3293 2006
597.96097—dc22   2005058592

The University Press of Florida is the scholarly publishing agency for the State
University System of Florida, comprising Florida A&M University, Florida Atlantic
University, Florida Gulf Coast University, Florida International University, Florida
State University, University of Central Florida, University of Florida, University of
North Florida, University of South Florida, and University of West Florida.

University Press of Florida
15 Northwest 15th Street
Gainesville, FL 32611-2079
http://www.upf.com

Contents

List of Species vi

Preface xv

1. Introduction 1
   How to Use This Book 2
   A Comment on Taxonomy 2
   Cautionary Notes 3
   Habitats 4
   Snakes in Captivity 4
   Snake Care 5
   About Snakes in General 6
   Key to the Families of North American Snakes 9

2. Blind Snakes 10

3. File Snakes 17

4. Boas and Pythons 20

5. Colubrid Snakes 28

6. Venomous Snakes 279

Glossary 329

References and Additional Reading 333

Index 337

Species

Nonvenomous Snakes

Blind Snakes

Slender Blind Snakes, family Leptotyphlopidae

1. Texas Blind Snake, *Leptotyphlops dulcis dulcis*
2. New Mexican Blind Snake, *Leptotyphlops dulcis dissectus*
3. Trans-Pecos Blind Snake, *Leptotyphlops humilis segregus*

Typical Blind Snakes, family Typhlopidae

4. Brahminy Blind Snake, *Ramphotyphlops braminus* (introduced)

File Snakes

File Snakes, family Acrochordidae

5. Javan File Snake, *Acrochordus javanicus* (introduced)

Boas and Pythons

Boas and Pythons, family Boidae

6. Colombian Boa Constrictor, *Boa constrictor imperator* (introduced)
7. Burmese Python, *Python molurus bivittatus* (introduced)
8. Ball Python, *Python regius* (introduced)

Colubrid Snakes

"Harmless" Snakes, family Colubridae

Racers, Whipsnakes, Crowned Snakes, and allies, subfamily Colubrinae

Racers, Whipsnakes, Patch-nosed Snakes, and Green Snakes

9. Northern Black Racer, *Coluber constrictor constrictor*
10. Buttermilk Racer, *Coluber constrictor anthicus*
11. Tan Racer, *Coluber constrictor etheridgei*
12. Eastern Yellow-bellied Racer, *Coluber constrictor flaviventris*
13. Blue Racer, *Coluber constrictor foxii*
14. Brown-chinned Racer, *Coluber constrictor helvigularis*
15. Black-masked Racer, *Coluber constrictor latrunculus*
16. Mexican Racer, *Coluber constrictor oaxaca*

17. Everglades Racer, *Coluber constrictor paludicola*

18. Southern Black Racer, *Coluber constrictor priapus*

19. Eastern Coachwhip Snake, *Masticophis flagellum flagellum*

20. Lined Coachwhip Snake, *Masticophis flagellum lineatulus*

21. Western Coachwhip Snake, *Masticophis flagellum testaceus*

22. Schott's Whipsnake, *Masticophis schotti schotti*

23. Ruthven's Whipsnake, *Masticophis schotti ruthveni*

24. Desert Striped Whipsnake, *Masticophis taeniatus taeniatus*

25. Central Texas Whipsnake, *Masticophis taeniatus girardi*

26. Eastern Indigo Snake, *Drymarchon corais couperi*

27. Texas Indigo Snake, *Drymarchon corais erebennus*

28. Northern Speckled Racer, *Drymobius margaritiferus margaritiferus*

29. Northern Rough Green Snake, *Opheodrys aestivus aestivus*

30. Florida Rough Green Snake, *Opheodrys aestivus carinatus*

31. Smooth Green Snake, *Opheodrys vernalis*

32. Mountain Patch-nosed Snake, *Salvadora grahamiae grahamiae*

33. Texas Patch-nosed Snake, *Salvadora grahamiae lineata*

34. Big Bend Patch-nosed Snake, *Salvadora hexalepis deserticola*

Hook-nosed Snakes and Ground Snakes

35. Tamaulipan Hook-nosed Snake, *Ficimia streckeri*

36. Chihuahuan Hook-nosed Snake, *Gyalopion canum*

37. Great Plains Ground Snake, *Sonora semiannulata semiannulata*

Black-headed Snakes and Crowned Snakes

38. Mexican Black-headed Snake, *Tantilla atriceps*

39. Southeastern Crowned Snake, *Tantilla coronata*

40. Trans-Pecos Black-headed Snake, *Tantilla cucullata*

41. Flat-headed Snake, *Tantilla gracilis*

42. Southwestern Black-headed Snake, *Tantilla hobartsmithi*

43. Plains Black-headed Snake, *Tantilla nigriceps*

44. Rimrock Crowned Snake, *Tantilla oolitica*

45. Peninsula Crowned Snake, *Tantilla relicta relicta*

46. Central Florida Crowned Snake, *Tantilla relicta neilli*

47. Coastal Dunes Crowned Snake, *Tantilla relicta pamlica*

Lyre Snake

48. Texas Lyre Snake, *Trimorphodon biscutatus vilkinsonii*

Black-striped Snakes, Night Snakes, Cat-eyed Snakes,
and Pinewoods Snakes, subfamily Dipsadinae

49. Tamaulipan Black-striped Snake, *Coniophanes imperialis imperialis*
50. Texas Night Snake, *Hypsiglena torquata jani*
51. Northern Cat-eyed Snake, *Leptodeira septentrionalis septentrionalis*
52. Pinewoods Snake, *Rhadinea flavilata*

Rat Snakes, Kingsnakes, Bullsnakes, and Allies, subfamily Lampropeltinae

Glossy Snakes

53. Kansas Glossy Snake, *Arizona elegans elegans*
54. Texas Glossy Snake, *Arizona elegans arenicola*
55. Painted Desert Glossy Snake, *Arizona elegans philipi*

Scarlet Snakes

56. Florida Scarlet Snake, *Cemophora coccinea coccinea*
57. Northern Scarlet Snake, *Cemophora coccinea copei*
58. Texas Scarlet Snake, *Cemophora coccinea lineri*

Rat Snakes

59. Trans-Pecos Rat Snake, *Bogertophis subocularis subocularis*
60. Baird's Rat Snake, *Elaphe bairdi*
61. Corn Snake, *Elaphe guttata guttata*
62. Great Plains Rat Snake, *Elaphe guttata emoryi*
63. Southwestern Rat Snake, *Elaphe guttata meahllmorum*
64. Black Rat Snake, *Elaphe obsoleta obsoleta*
65. Texas Rat Snake, *Elaphe obsoleta lindheimeri*
66. Yellow Rat Snake, *Elaphe obsoleta quadrivittata*
67. Everglades Rat Snake, *Elaphe obsoleta rossalleni*
68. Gray Rat Snake, *Elaphe obsoleta spiloides*
69. Gulf Hammock Rat Snake, *Elaphe obsoleta quadrivittata* x *E. o. spiloides*
70. Western Fox Snake, *Elaphe vulpina vulpina*
71. Eastern Fox Snake, *Elaphe vulpina gloydi*

Kingsnakes and Milksnakes

72. Gray-banded Kingsnake, *Lampropeltis alterna*
73. Prairie Kingsnake, *Lampropeltis calligaster calligaster*
74. South Florida Mole Kingsnake, *Lampropeltis calligaster occipitolineata*
75. Mole Kingsnake, *Lampropeltis calligaster rhombomaculata*
76. Eastern Kingsnake, *Lampropeltis getula getula*

77. California Kingsnake, *Lampropeltis getula californiae* (introduced)
78. Florida Kingsnake, *Lampropeltis getula floridana*
79. Peninsula Kingsnake, *Lampropeltis getula floridana* x *L. g. getula*
80. Apalachicola Lowland Kingsnake, *Lampropeltis getula* ssp.
81. Speckled Kingsnake, *Lampropeltis getula holbrooki*
82. Eastern Black Kingsnake, *Lampropeltis getula nigra*
83. Desert Kingsnake, *Lampropeltis getula splendida*
84. Outer Banks Kingsnake, *Lampropeltis getula sticticeps*
85. Eastern Milksnake, *Lampropeltis triangulum triangulum*
86. Louisiana Milksnake, *Lampropeltis triangulum amaura*
87. Mexican Milksnake, *Lampropeltis triangulum annulata*
88. New Mexican Milksnake, *Lampropeltis triangulum celaenops*
89. Scarlet Kingsnake, *Lampropeltis triangulum elapsoides*
90. Central Plains Milksnake, *Lampropeltis triangulum gentilis*
91. Pale Milksnake, *Lampropeltis triangulum multistriata*
92. Red Milksnake, *Lampropeltis triangulum syspila*
93. Coastal Plains Milksnake, *Lampropeltis triangulum* ssp.

Gopher Snakes, Pine Snakes, and Bullsnakes
94. Sonoran Gopher Snake, *Pituophis catenifer affinis*
95. Louisiana Pine Snake, *Pituophis catenifer ruthveni*
96. Bullsnake, *Pituophis catenifer sayi*
97. Northern Pine Snake, *Pituophis melanoleucus melanoleucus*
98. Black Pine Snake, *Pituophis melanoleucus lodingi*
99. Florida Pine Snake, *Pituophis melanoleucus mugitus*
100. Texas Long-nosed Snake, *Rhinocheilus lecontei tessellatus*
101. Short-tailed Snake, *Stilosoma extenuatum*

Water Snakes, Garter Snakes, Earth Snakes, and allied species, subfamily Natricinae

Kirtland's Snake and Water Snakes
102. Kirtland's Snake, *Clonophis kirtlandi*
103. Gulf Salt Marsh Snake, *Nerodia clarkii clarkii*
104. Mangrove Salt Marsh Snake, *Nerodia clarkii compressicauda*
105. Atlantic Salt Marsh Snake, *Nerodia clarkii taeniata*
106. Mississippi Green Water Snake, *Nerodia cyclopion*
107. Red-bellied Water Snake, *Nerodia erythrogaster erythrogaster*
108. Yellow-bellied Water Snake, *Nerodia erythrogaster flavigaster*

109. Copper-bellied Water Snake, *Nerodia erythrogaster neglecta*
110. Blotched Water Snake, *Nerodia erythrogaster transversa*
111. Banded Water Snake, *Nerodia fasciata fasciata*
112. Broad-banded Water Snake, *Nerodia fasciata confluens*
113. Florida Water Snake, *Nerodia fasciata pictiventris*
114. Florida Green Water Snake, *Nerodia floridana*
115. Brazos Water Snake, *Nerodia harteri harteri*
116. Concho Water Snake, *Nerodia harteri paucimaculata*
117. Diamondbacked Water Snake, *Nerodia rhombifer rhombifer*
118. Northern Water Snake, *Nerodia sipedon sipedon*
119. Lake Erie Water Snake, *Nerodia sipedon insularum*
120. Midland Water Snake, *Nerodia sipedon pleuralis*
121. Carolina Water (salt marsh) Snake, *Nerodia sipedon williamengelsi*
122. Brown Water Snake, *Nerodia taxispilota*

Crayfish Snakes

123. Striped Crayfish Snake, *Regina alleni*
124. Graham's Crayfish Snake, *Regina grahamii*
125. Glossy Crayfish Snake, *Regina rigida rigida*
126. Delta Crayfish Snake, *Regina rigida deltae*
127. Gulf Crayfish Snake, *Regina rigida sinicola*
128. Queen Snake, *Regina septemvittata*

Swamp Snakes

129. North Florida Swamp Snake, *Seminatrix pygaea pygaea*
130. South Florida Swamp Snake, *Seminatrix pygaea cyclas*
131. Carolina Swamp Snake, *Seminatrix pygaea paludis*

Brown Snakes and Red-bellied Snakes

132. Northern Brown Snake, *Storeria dekayi dekayi*
133. Marsh Brown Snake, *Storeria dekayi limnetes*
134. Texas Brown Snake, *Storeria dekayi texana*
135. Florida Brown Snake, *Storeria dekayi victa*
136. Midland Brown Snake, *Storeria dekayi wrightorum*
137. Northern Red-bellied Snake, *Storeria occipitomaculata occipitomaculata*
138. Florida Red-bellied Snake, *Storeria occipitomaculata obscura*
139. Black Hills Red-bellied Snake, *Storeria occipitomaculata pahasapae*

Garter Snakes, Ribbon Snakes, and Lined Snakes

140. Short-headed Garter Snake, *Thamnophis brachystoma*
141. Butler's Garter Snake, *Thamnophis butleri*

142. Western Black-necked Garter Snake, *Thamnophis cyrtopsis cyrtopsis*
143. Eastern Black-necked Garter Snake, *Thamnophis cyrtopsis ocellatus*
144. Wandering Garter Snake, *Thamnophis elegans vagrans*
145. Checkered Garter Snake, *Thamnophis marcianus marcianus*
146. Orange-striped Ribbon Snake, *Thamnophis proximus proximus*
147. Aridland Ribbon Snake, *Thamnophis proximus diabolicus*
148. Gulf Coast Ribbon Snake, *Thamnophis proximus orarius*
149. Red-striped Ribbon Snake, *Thamnophis proximus rubrilineatus*
150. Plains Garter Snake, *Thamnophis radix*
151. Eastern Ribbon Snake, *Thamnophis sauritus sauritus*
152. Blue-striped Ribbon Snake, *Thamnophis sauritus nitae*
153. Peninsula Ribbon Snake, *Thamnophis sauritus sackenii*
154. Northern Ribbon Snake, *Thamnophis sauritus septentrionalis*
155. Eastern Garter Snake, *Thamnophis sirtalis sirtalis*
156. Texas Garter Snake, *Thamnophis sirtalis annectens*
157. New Mexican Garter Snake, *Thamnophis sirtalis dorsalis*
158. Maritime Garter Snake, *Thamnophis sirtalis pallidulus*
159. Red-sided Garter Snake, *Thamnophis sirtalis parietalis*
160. Chicago Garter Snake, *Thamnophis sirtalis semifasciatus*
161. Blue-striped Garter Snake, *Thamnophis sirtalis similis*
162. Lined Snake, *Tropidoclonion lineatum*

Earth Snakes
163. Rough Earth Snake, *Virginia striatula*
164. Eastern Earth Snake, *Virginia valeriae valeriae*
165. Western Earth Snake, *Virginia valeriae elegans*
166. Mountain Earth Snake, *Virginia valeriae pulchra*

Odd-toothed Snakes, subfamily Xenodontinae

Worm Snakes
167. Eastern Worm Snake, *Carphophis amoenus amoenus*
168. Midwest Worm Snake, *Carphophis amoenus helenae*
169. Western Worm Snake, *Carphophis amoenus vermis*

Ring-necked Snakes
170. Southern Ring-necked Snake, *Diadophis punctatus punctatus*
171. Key Ring-necked Snake, *Diadophis punctatus acricus*
172. Prairie Ring-necked Snake, *Diadophis punctatus arnyi*
173. Northern Ring-necked Snake, *Diadophis punctatus edwardsi*

174. Regal Ring-necked Snake, *Diadophis punctatus regalis*
175. Mississippi Ring-necked Snake, *Diadophis punctatus stictogenys*

Mud Snakes and Rainbow Snakes

176. Eastern Mud Snake, *Farancia abacura abacura*
177. Western Mud Snake, *Farancia abacura reinwardtii*
178. Common Rainbow Snake, *Farancia erytrogramma erytrogramma*
179. South Florida Rainbow Snake, *Farancia erytrogramma seminola*

Hog-nosed Snakes

180. Plains Hog-nosed Snake, *Heterodon nasicus nasicus*
181. Dusty Hog-nosed Snake, *Heterodon nasicus gloydi*
182. Mexican Hog-nosed Snake, *Heterodon nasicus kennerlyi*
183. Eastern Hog-nosed Snake, *Heterodon platirhinos*
184. Southern Hog-Nosed Snake, *Heterodon simus*

Venomous Snakes

Coral Snakes

Coral Snakes, family Elapidae

185. Eastern Coral Snake, *Micrurus fulvius*
186. Texas Coral Snake, *Micrurus tener*

Pit Vipers

Copperheads, Cottonmouths, and Rattlesnakes, family Viperidae

Copperheads and Cottonmouths

187. Southern Copperhead, *Agkistrodon contortrix contortrix*
188. Broad-banded Copperhead, *Agkistrodon contortrix laticinctus*
189. Northern Copperhead, *Agkistrodon contortrix mokasen*
190. Osage Copperhead, *Agkistrodon contortrix phaeogaster*
191. Trans-Pecos Copperhead, *Agkistrodon contortrix pictigaster*
192. Eastern Cottonmouth, *Agkistrodon piscivorus piscivorus*
193. Florida Cottonmouth, *Agkistrodon piscivorus conanti*
194. Western Cottonmouth, *Agkistrodon piscivorus leucostoma*

Rattlesnakes

195. Eastern Diamondbacked Rattlesnake, *Crotalus adamanteus*
196. Western Diamondbacked Rattlesnake, *Crotalus atrox*
197. Timber Rattlesnake, *Crotalus horridus horridus*
198. Canebrake Rattlesnake, *Crotalus horridus atricaudatus*
199. Mottled Rock Rattlesnake, *Crotalus lepidus lepidus*
200. Banded Rock Rattlesnake, *Crotalus lepidus klauberi*
201. Northern Black-tailed Rattlesnake, *Crotalus molossus molossus*
202. Mojave Rattlesnake, *Crotalus scutulatus scutulatus*
203. Prairie Rattlesnake, *Crotalus viridis viridis*
204. Eastern Massasauga, *Sistrurus catenatus catenatus*
205. Desert Massasauga, *Sistrurus catenatus edwardsi*
206. Western Massasauga, *Sistrurus catenatus tergeminus*
207. Carolina Pygmy Rattlesnake, *Sistrurus miliarius miliarius*
208. Dusky Pygmy Rattlesnake, *Sistrurus miliarius barbouri*
209. Western Pygmy Rattlesnake, *Sistrurus miliarius streckeri*

Preface

Today, as interest in and knowledge of our reptile fauna increase, so, too, do the challenges to the long-term continued existence of many of these interesting creatures. Pressures such as climatic change, habitat destruction (including fragmentation), roadkill, collection for the pet trade, competition from introduced species, and wanton killing impact snakes and other reptiles.

On the plus side, snakes are *hot*, as a topic. Interpretive programs and study units on snakes are part of elementary, middle, and high school curricula, and undergraduate and postgraduate college courses pertaining specifically to herpetology are proliferating. Ecotours featuring reptile fauna are a short airline flight away.

The snakes in this book are found in eastern and central North America, north of Mexico. The westernmost limit is western Texas, western North Dakota, and western Manitoba. Most of these 209 species are native, but almost half a dozen species are introduced and established. Nationally, there are more than sixty species of introduced and established reptiles and amphibians. Most of the alien snake species are of minimal impact, but one, the Burmese python in Florida, is of significant concern.

The Ecological Impact of Man on the South Florida Herpetofauna, written by Larry David Wilson and Louis Porras, published in 1983, detailed the presence and effects of the twenty-five species of alien herpetofauna then established in the state. Although now dated (there are now more than sixty introduced and established species of herpetofauna, dozens of species of birds, and a few mammal species in the state), the book remains relevant. The term "ecocollapse" was used to describe what they felt were ever-worsening (and, twenty-plus years later, still unresolved) environmental and ecological conditions in Florida, south of Lake Okeechobee.

Snakes are not always evenly distributed even within their delineated ranges. Some are of montane distribution, occurring only between certain elevations on certain mountains, some dwell in acidic bogs or in arid habitats, and still others may predominate in wetlands or even in suburban settings. A specific reptile may be abundant in one state or province yet so rare in a neighboring area that it is considered endangered. Some snakes are actually common throughout their range.

Because of the varying and ever changing state and provincial laws, we have not attempted in most cases to advise readers of the legal status of a snake, but may occasionally make general statements on legalities. We urge that before you collect a particular species you check both federal and state laws to determine its legal status. You may need a permit (usually issued only for research purposes) before seeking a specific reptile.

We now invite you to join us as we tour mountains, deserts, woodlands, and backyards, looking beneath rocks, in grass tussocks, and in roadside dumps for the snakes of eastern and central North America north of Mexico.

Acknowledgments

The success of a publication such as this is due largely to the efforts and generosity of colleagues and friends. With this in mind, we gratefully acknowledge the comments and concerns of Kevin Enge, Richard Franz, Paul Moler, and R. Wayne Van Devender.

We are grateful to David Auth, who allowed us access to the study collection of the Florida Museum of Natural History, where we were able to see the enigmatic South Florida rainbow snake firsthand.

Collette Adams, Ray E. Ashton, Chris Bednarski, Bill and Marcia Brant, Karin Burns, Dennis Cathcart, Scott Cushnir, Norm Damm, John Decker, C. Kenneth Dodd, James Duquesnel, Harry Greene, Lance Jarzynka, Dennie Miller, Norm Nunley, Justin Garza, Billy Griswold, Jim Harding, Terry Hibbitts, Toby Hibbitts, Troy Hibbitts, Joe Hiduke, Steven Johnson, Mark Kenderdine, F. Wayne King, Ken King, Kenney Krysko, John Lewis, Mark Ludlow, Dave Manke, John McGonigal, Carl May, Mike Manfredi, Barry Mansell, Brian Mealey, Flavio Morrissey, David Nelson, Sandy Oldershaw, Regis Opferman, Charlie Painter, Greta Parks, Dan Pearson, Andy Price, Mike Price, Gus Rentfro, Alan Resetar, Jeff Schofield, Dan Scolaro, Don Sias, Brian Smith, Mike Smith, Tom Tyning, Rick Van Dyke, Jerry Walls, Frank Weed, Pete Wilson, Larry Wood, and Maria and Kenny Wray either joined us, or allowed us to join them, in the field, provided specimens for us to photograph, or, in some cases, provided the photographs themselves. Thanks to all.

Bill Love, Rob MacInnes, Chris McQuade, and Eric Thiss made possible some of the photographs in this volume.

Kevin Enge and Walter Meshaka shared with us their encyclopedic knowledge of the introduced reptiles and amphibians of Florida and supplied the field data that allowed us to find and photograph many of these herpetological interlopers in the wild.

1

Introduction

The United States is endowed with a remarkable diversity of reptiles. East of the somewhat arbitrary line now used to demarcate eastern and central North America—from western Texas northward to and including Canada's Keewatin region—there are more than 400 species and subspecies of reptiles. Of this number, about half are snakes. These range from tiny eyeless burrowing snakes to six-foot rattlesnakes and, now, fifteen-foot Burmese pythons.

At least four snake species have been introduced into the fauna of the United States. The most spectacular of these four is the Burmese python. The smallest species is the tiny burrowing Brahminy blind snake. Since these introductions are rather recent, we don't yet know what, if any, pressures these creatures are placing on populations of native wildlife.

The native snakes in eastern and central North America vary widely in size as well. At one extreme are the tiny brown snakes and ring-necked snakes and at the other are the large indigo and rat snakes—but here the maximum size is nine feet.

Generally, snakes are categorized as diurnal, nocturnal, or crepuscular. Actually, most are active during more than one of these times and their activity patterns may change seasonally or even with short-term climatic changes.

Although populations of most of our snakes do not long persist in areas of urban and suburban sprawl, a few, such as corn snakes and garter snakes, are eminently successful in these disturbed habitats.

Many biologists, no matter their affiliation, feel that legal protection should be offered to more species, although legal protection alone won't do the job. Assuring that these creatures remain for us and our descendants to view and appreciate in the wild will take concerted effort.

Whether we are researchers or herpetoculturists, or merely have an interest in the creatures with whom we share our world, it is time for us to join forces and promote the conservation of these interesting, beneficial, and highly specialized animals.

We hope that our comments in this identification guide will help you to better understand and appreciate the intricate lifestyles of our eastern snakes.

How to Use This Book

In this volume we discuss the 209 species and subspecies of snakes that you may see in eastern and central North America, north of Mexico. Some are easily identifiable, being so different from all other species that it would be difficult to mistake them, and most have at least one feature by which they may be most easily recognized.

Sometimes identification can be a bit of a challenge—many snakes have two or even more very different color phases and the interbreeding of two forms can create progeny that are confusing in appearance and difficult to identify. Subtle characteristics such as the positioning of blotches or stripes are often important identifying factors. Whether the scales are keeled or smooth, dull or shiny, and whether the creature's pupils are vertically elliptical or round are also important. (Keep in mind, though, that if you are close enough to see the pupil shape of a large venomous snake, you are probably too close for your own safety!) Certain characteristics are not easily determined unless the snake is in hand—and there are some reptiles that you just should not have in hand, either for your own safety or for legal reasons.

There are times when a snake is so aberrant that no normally pertinent factor will help you with identification. Albinism or scalelessness can obliterate patterns and scale characteristics (but the chance of actually finding an albino or scaleless snake in the wild is pretty slim because natural selection weeds out highly visible albinos and snakes with soft, scaleless skin). When in doubt, one must depend on range and "gestalt" for identification.

We have numbered all species and subspecies in the table of contents, and these coincide with the numbers assigned in both text and photographs. All species and subspecies and several of the naturally occurring intergrades within our range are pictured. There is no other field guide currently available that provides this important identification tool.

A Comment on Taxonomy

The science of classification, of naming species, is called taxonomy. Scientific names are of Latin or Greek derivation. They can be binomial (two names) or trinomial (three names). A trinomial indicates that a species has more than one recognized subspecies or race.

There is a tendency today to use molecular data (genetic information) to identify separate species among very similar appearing creatures. With this comes a tendency to discount subspecies. Before the development of molecular data, these very similar appearing creatures were classified as subspecies, and field markings used to distinguish one from another. We have opted to list and discuss the eastern and central snakes in a traditional manner, divided into families, subfamilies, genera, species, and subspecies. We have also elected to use the most conservative taxonomy but have made mention of proposed nomenclature changes. Time and usage will determine which will prevail.

As in any other discipline, there are diverging beliefs, techniques, and applications and the proponents of one often vociferously decry the suggestions and conclusions of the other. Because traditional systematics has "worked well" over the years, and because we feel that a field guide is not the proper forum for arguing taxonomic principles, we continue to take a comfortable and conservative approach in these pages. Almost certainly some names will change in future editions.

Wherever we felt it possible, both the common and scientific names used in this book are those suggested in *Scientific and Standard Common Names of Amphibians and Reptiles of North America North of Mexico, with Comments Regarding Confidence in Our Understanding* (Crother, 2000).

Cautionary Notes

Do not approach venomous snakes. The vast majority of snakebites in the United States and Canada happen when an observer attempts to kill, move, or otherwise confront a venomous snake. A smaller percentage of bites happen when a hiker or rock climber inadvertently steps on or touches a venomous snake.

Of the reptiles of eastern and central North America, only snakes are venomous. Because of the varying techniques for treating snake envenomation and the potential for anaphylactic reactions to some treatments, we have elected to say only that consultation with a qualified physician should immediately be sought following any venomous snakebite.

Nonvenomous snakes will bite if frightened. There is a chance of contamination by pathogens from any bite. Abrasions or bites should be thoroughly cleansed; if the damage is serious, professional medical attention should be sought.

Habitats

To find a particular species or subspecies of snake, you must look first within its range, and next in its proper habitat. Many snakes are habitat generalists, but others require a specific niche (microhabitat) within their habitat. Some may be found only in the vicinity of freshwater ponds, others may dwell among mangroves or salt marsh grasses. Some are restricted to lowland habitats while others occur only in montane (mountainous) regions. Trash piles, building rubble, discarded boards, and roofing tin are favored hiding areas. Specific snake species may be found in woodlands, pastures, deserts, old fields, or riparian situations.

Despite ongoing habitat modifications, there are still vast stretches suitable for snakes in the United States and Canada. In some cases, habitat modification has worked in favor of certain snakes. For example, pastures are ideal habitats for racers, and stock watering tanks now provide moisture to desert species.

Snakes in Captivity

This section is intended only as an overview. We have included a few helpful snake care titles in the reference section.

Before seeking snakes in the wild, or keeping them in captivity, you are urged to check the legal status of these animals with the game and fish commission in each jurisdiction as well as with the U.S. Fish and Wildlife Service. Some species now have protected or regulated status and should not collected, or even molested, without specific permits. Up-to-date regulations are readily available online, on the Web sites of the regulatory agencies.

The keeping of reptiles and amphibians in captivity, snakes in particular, has become a popular hobby. Providing the animals and the necessary equipment and supplies to an eager populace is a multimillion-dollar business.

Collecting snakes from the wild for the pet trade is less acceptable today than in bygone years, yet it is entirely legal in many areas. Even if certain jurisdictions do not allow commercial collecting, they often do allow the collecting of one to several snakes of various kinds for personal use. The sale of captive-bred native snakes is usually allowed and should be supported whenever possible. Many states do not protect introduced exotic reptiles.

Some snakes are easily kept, requiring only a secure terrarium, food, water, and periodic cleaning to thrive as captives, and their lifespan may exceed twenty years. Other species (and an occasional individual of normally easily kept kinds) may prove very difficult captives. Both aspects must be considered when you're thinking of acquiring a snake.

There was a time when the short-term keeping of snakes (and other reptiles) was an accepted practice. A snake would be found, picked up, and brought home by an enthusiast, kept for a period of time, then released. However, it has been learned that in many cases, unless snakes are released in almost the exact spot from which they were collected, they are not able to reacclimate and their survival rate is low. Likewise, a snake purchased from a dealer should never be turned loose. It isn't fair to the snake or to the wild populations, because there is always the chance of introducing a pathogen contracted from you or from captive reptiles.

If you find you are not prepared to undergo a long-term commitment, don't collect snakes from the wild. Finding and photographing them is rewarding and infinitely kinder than capturing and then neglecting and abandoning them.

Snake Care

Suitable terrarium/cage conditions for one species of snake may not be adequate for another species. Research the needs of your potential pet carefully. Provide a terrarium or cage that is as large as possible (some states have laws regulating minimum cage size in comparison to snake size).

Although there are now many types of cages available specifically for reptiles, we usually use covered aquaria. Choose a tank that's large enough, and make certain the cover fits tightly.

The availability of fresh water to a snake is of paramount importance. Fortunately, most snakes readily drink from a bowl. However, the cage itself must be dry to discourage bacterial infections that attack snakes' skin.

In general, snakes need to be kept warm to remain active and to allow for natural bodily functions. Thermal gradients should be provided in the terrarium; i.e., warm on one end, cooler on the other. Heat can be provided through the use of overhead bulbs. The suggested basking temperature for most is in the high 80s to mid 90s F. Be sure the cool end of the cage is at least 10°F cooler. Never place a glass terrarium or cage in direct sunlight. Glass intensifies and concentrates the heat and the elevated temperatures will quickly kill even the most heat-tolerant reptiles.

Some snakes are difficult to keep because of their specialized diet. Among these are scarlet snakes (reptile eggs), ground and earth snakes (burrowing insects), crowned and black-headed snakes (insect larvae and centipedes), short-tailed snake (a protected species that feeds on crowned snakes), and mud and rainbow snakes (aquatic salamanders and eels respectively).

The venomous snakes and large boas and pythons should not be obtained/collected and kept unless you are experienced and licensed (where necessary) and it is legal to keep them in your municipality.

Most of the rat snakes, kingsnakes, and pine snakes (rodent eaters) are pretty, hardy, and easily kept; if a diet of amphibians and reptiles is provided, racers and whipsnakes are also easy to keep. If you have a ready supply of worms, fish, and frogs available, the garter snakes, ribbon snakes, and water snakes are equally easy to keep.

About Snakes in General

Snakes are instantly recognizable to almost everyone. They are virtually legless (a few retain remnants of rear limbs in the form of a keratinized spur on each side of the anal opening). Those with functional, external eyes have them covered by a clear spectacle (the brille) or by a modified clear scale. They are found in most areas of the world except Arctic and Antarctic regions, Iceland, Ireland, and New Zealand.

There are more than 2,400 species of snakes in the world, of which about 450 species are venomous in the most traditional sense of the word. There are many other species that produce salivary toxins but are classified as harmless nonetheless.

Because of the diversity of snakes' habitats, the two states within the scope of our coverage having the greatest number and diversity of snake species are Florida and Texas.

The snakes of the eastern and central states, including established non-native (exotic) species, are contained in the following families:

Blind Snakes, family Typhlopidae (1 species)
Slender Blind Snakes, family Leptotyphlopidae (2 species)
File Snakes, family Acrochordidae (1 species)
Boas and Pythons, family Boidae (3 species)
Typical Harmless Snakes, family Colubridae (82 species)
Cobra Allies (Coral Snakes), family Elapidae (1 species)
Pit Vipers, family Viperidae (11 species)

These families comprise a total of 206 species and subspecies.

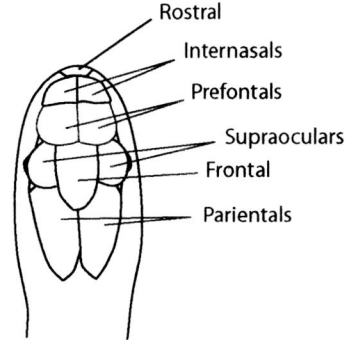

Figure 1.01. Top of snake head. Illustration by Patricia Bartlett.

1. Rostral
2. Nasals
3. Loreal
4. Preoculars
5. Supraocular
6. Postoculars
7. Temporals
8. Parietals
9. Upper labials
10. Lower labials

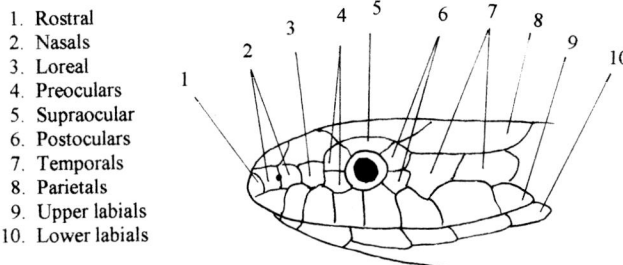

Figure 1.02. Side of snake head. Illustration by Patricia Bartlett.

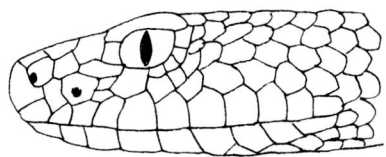

Figure 1.03. Pit viper head. Illustration by Patricia Bartlett.

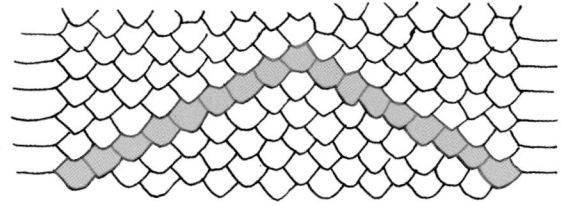

Figure 1.04. Counting scale rows. Illustration by Patricia Bartlett.

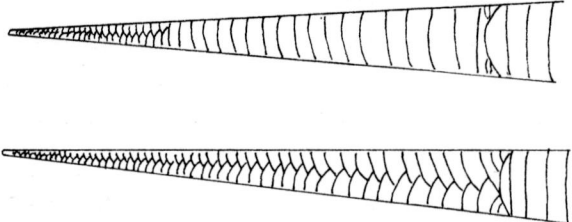

Figure 1.05. Subcaudals. Illustration by Patricia Bartlett.

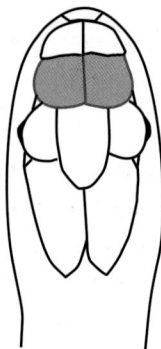

Figure 1.06. Snake head showing 2 prefrontals. Illustration by Patricia Bartlett.

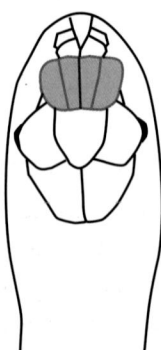

Figure 1.07. Snake head showing 4 prefrontals. Illustration by Patricia Bartlett.

Key to the Families of North American Snakes

1a. Size, tiny. No enlarged scales on belly. No functional external eyes. Tail blunt, spine tipped. 2

1b. Enlarged scales on belly, functional eyes. 3

2a. Teeth only in lower jaw. 14 scale rows. . . . .
. . . . **Leptotyphlopidae: Slender Blind Snakes**

2b. Teeth only in upper jaw. 16 scale rows. . . . .
. . . . **Typhlopidae: Typical Blind Snakes**

3a. Size small to immense. Body scales small. Ventral plates present. Crown (top of head) scales very small (fragmented). 4

3b. Body scales large. 5

4a. Pupils elliptical. Labial (lip) pits present. No movable fangs at front of upper jaw. . . . . **Boidae: Pythons, Boas**

4b. No labial pits. 5

5a. Body scales large. Pupils elliptical. Loreal pits present. Crown scales very small (fragmented) or nonfragmented. Movable fangs at front of upper jaw (venomous) . . . . **Viperidae, in part**
(Crotalinae: Pit Vipers, Rattlesnakes, Cottonmouths, Copperheads)

5b. Loreal pits absent. 6

6a. Body scales large. Ventral plates large. Pupils elliptical or round. Crown scales unfragmented. No fangs at front of upper jaw. Color and pattern variable (if ringed with black, red, and yellow or white, see 7) . . . .
. . . . **Colubridae**

6b. As above but patterned with red, yellow (or white), and black rings.

7a. Yellow and red rings not next to each other . . . . **Colubridae**

7b. As above but with red and yellow rings touching 8

8a. Short, permanently erect fangs in front of mouth. No loreal pit. Traffic light caution colors of red and yellow touching (yellow may be lacking in Upper Florida Keys). Head small and somewhat flattened (venomous) . .
. . . . **Elapidae (Micrurinae: Coral Snakes)**

8b. Color not as above. Loreal pits present. Movable fangs at front of upper jaw (venomous) . . . .
. . . . **Viperidae, in part (Crotalinae: Pygmy Rattlesnakes, Massasaugas)**

2

Blind Snakes

Slender Blind Snakes and Typical Blind Snakes, families Leptotyphlopidae and Typhlopidae

Skull characteristics separate the snakes in these two externally similar families. The slender blind snakes (Leptotyphlopidae) have teeth only in the lower jaw, and the typical blind snakes (Typhlopidae) lack teeth in the lower jaw. The word "slender" is not used in their common names.

These are primarily tropical and Southern Hemisphere snakes. They are poorly represented in the United States. The two species of slender blind snakes are native to North America, and occur from Texas and Kansas westward. The single species of typical blind snake in the United States, the Brahminy blind snake, is of Asian origin. It has been established in South Florida since at least the early 1980s, and in the lower Rio Grande valley of Texas since the early 1990s. Because it is commonly found in the soil in flowerpots, which is how it is introduced to an area, it is also known as the flowerpot snake.

All blind snakes are nonvenomous and persistent burrowers. They do not have functional eyes and the belly scales are not enlarged. The lower jaw is noticeably countersunk, an ideal adaptation for any burrowing species. These snakes feed upon burrowing insects and their larvae, and all seem especially partial to the pupae and larvae of ants and subterranean termites. Those in the United States are diminutive, the largest rarely attaining a foot in length.

Within a given genus, the arrangement of scales on the head is the most effective species designator, but actually being able to see the scales will require good light and, in most cases, a magnifying glass. Check the range maps for additional help.

Slender Blind Snakes, family Leptotyphlopidae

Slender Blind Snakes, genus *Leptotyphlops*

This is a large genus of predominantly small snakes that occur in many of the world's warmer regions. The eyes are not functional and there are teeth only in the lower jaw, although the jaws are far too small for the teeth to be seen easily.

These snakes are inhabitants of arid and semiarid lands and are often found beneath flat rocks or surface debris. They are most common where some ground moisture is present, and may be active after heavy rains.

When picked up, a blind snake squirms animatedly, may open its mouth, and smears the captor with feces and urates from the cloaca. A snake that is attacked by ants tips the body scales forward, writhes, and smears itself with its cloacal contents, apparently as a deterrent.

A single small clutch (2–7) of eggs is laid annually. Females apparently remain with the clutch throughout the 45–55 days of incubation. Communal nesting has been documented.

Notes on the blind snakes

Lightning flashed from cloud to ground as the tiny snake wriggled animatedly across the pavement. Until the clouds from the impending storm obscured the heavens, it had been a quiet night in west Texas spent commenting on meteorite activity, puzzling over the enigmatic and notorious lights seen in the sky east of Marfa, and looking for snakes. Snake activity had been curiously nil until the little New Mexico blind snake made its appearance. We left the car and knelt to photograph the snake as the first spattering drops of rain hit the pavement. The air, which only moments earlier had smelled dusty and heavy, suddenly smelled clean, clear, and redolent with the fresh scent of creosote bush. Snake activity or not, this was a wonderful late summer night.

I couldn't help but think how different the habitat of the western blind snake is from that of the introduced Brahminy blind snake. The snake at my feet is a resident of rock-strewn yielding sands in semi-arid regions, while the Brahminy blind snake prefers life in the humid, warm, and often wet (but well-drained) soils of southern Florida and south Texas. There it is an unnoticed resident of manicured gardens and even the soil of potted plants. It is probably via soil in potted plants that this minuscule snake is expanding its range as unwary humans transport pots far beyond their origins.

When I returned to Florida I sought out the Brahminy blind snake. A friend took me to a debris-filled field in central Florida and I soon realized exactly how common—and how agile—they are. Within just a few minutes, we fleetingly saw dozens of the tiny serpents beneath wet, discarded carpets, retreating into their tiny tunnels as we rolled the carpets back.

Although this central Florida location was the furthest north Brahminy blind snakes were known in the mid-1990s, they have continued their march northward and westward in potted plants shipped from southern Florida. It is unlikely that they will become established in more northerly climes, but we are uncertain about the actual level of tolerance these snakes have for the cold.

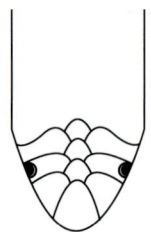

Figure 2.01. Blind Snake.
Illustration by Dale Johnson.

1. Texas Blind Snake

Leptotyphlops dulcis dulcis

Abundance/Range: This is a common snake that is easily overlooked. It ranges over most of central Texas as well as southwestern Oklahoma and adjacent Mexico.

Habitat: The Texas blind snake may be encountered in remote prairie and rangeland habitats as well as in urban settings. It is an excellent burrower, but may be found beneath all types of moisture-retaining surface debris. Heavy, burrow-flooding rains bring this little earthworm look-alike to the surface.

Size: These snakes are adult at 5–9 inches in length. The largest specimen yet found was slightly less than 11 inches. Hatchlings are about 2¾ inches long.

Identifying features: It is not difficult to identify these tiny serpents as blind snakes, but differentiating the species and subspecies is a more daunting task.

The Texas blind snake is a small reddish, purplish, tan, or brown snake with a lighter belly. When the snake nears ecdysis, or gets ready to shed its skin, the color changes to grayish and is almost opalescent. These snakes are often mistaken for earthworms, but lack annuli (body rings).

The eyes are barely visible beneath a large scale (the ocular scale). The ocular

1. Texas Blind Snake

Texas
New Mexican

scales are separated on the top of the head by 3 scales and there is only a single supralabial (upper lip) scale. The tail is bluntly rounded, but tipped with a spine-like scale. There are no enlarged belly scales.

Similar species: The very similar Trans-Pecos blind snake, *L. humilis segregus*, has only a single scale separating the ocular scales on the top of the head.

Additional subspecies

2. The New Mexico Blind Snake, *Leptotyphlops dulcis dissectus*, differs only in having two supralabial scales. This race is found westward from west Texas to southeastern Arizona (and adjacent Mexico), southwestern Kansas, western Oklahoma, and immediately adjacent southeastern Colorado, and in a disjunct area of central New Mexico.

2. New Mexico Blind Snake

3. Trans-Pecos Blind Snake

Leptotyphlops humilis segregus

Abundance/Range: This is a common snake that, because of its burrowing propensities, is seldom seen. It is found from the eastern Big Bend of Texas westward to southeastern Arizona. It is also found in adjacent Mexico.

Habitat: Canyonlands, grasslands, semiarid and arid scrublands, and desert are the habitats of this snake. It may burrow quite deeply during times of drought, but when and where a little surface moisture is present, the Trans-Pecos blind snake often rests on the ground beneath rocks and other surface debris.

Size: The largest of the blind snakes of the United States, this species tops out at about 13 inches. However, most examples are 2 or 3 inches shorter. Hatchlings are about 3 inches in length.

Identifying features: This is a blunt-tailed, cylindrical snake, with eyes obscured by the ocular scales. It has only one scale between the ocular scales on the top of the head. It is purplish brown to brown dorsally with a lighter belly.

Similar species: Few snakes can be confused with the blind snakes. Only the two subspecies of the Texas blind snake occur within the range of the Trans-Pecos blind snake. Both races of the Texas blind snake, *L. dulcis*, have the ocular scales separated by three small scales on the top of the head.

3. Trans-Pecos Blind Snake. Photo by R. Wayne Van Devender

Typical Blind Snakes, family Typhlopidae

Typical Blind Snakes, genus *Ramphotyphlops*

A single species in the genus *Ramphotyphlops* has become established in the southeastern United States.

Very similar to the slender blind snakes in external appearance, the snakes of this genus *lack* teeth in the lower jaw. The sole species found in the United States probably arrived here in potted plants. Species in this snake family occur naturally as close as the Bahamas.

The Brahminy blind snake is a unisexual, all female, parthenogenetic, normally oviparous species. Records exist of this species occasionally producing live young.

4. Brahminy Blind (Flowerpot) Snake

Ramphotyphlops braminus

Abundance/Range: Because it is inadvertently moved about in pots of nursery stock, this snake is reported from several new areas of the country each year. Among the latest are coastal Georgia and southern Alabama. Because it is sensitive to cold, it is probably able to survive in the wild only in subtropical South Florida and the lower Rio Grande valley. It is an abundant but seldom seen snake. Now found virtually worldwide in subtropical and tropical regions, this snake is of Asian origin.

Habitat: Look for this snake under stepping stones in gardens and lawns, beneath piles of vegetational debris, or under any type of moisture-retaining rubble. This snake probably entered the continental United States in nursery stock. It rapidly became established in rainy subtropical southern Florida and was found beneath discarded carpets in the lower Rio Grande valley of Texas in the early 1990s.

4. Brahminy Blind Snake

Size: This smallest of the blind snakes is adult at 3–5 inches in length. It is also proportionately slender, a large adult being about the diameter of a knitting needle. Hatchlings/neonates are about 2¼ inches long.

Identifying features: Normally a shiny black above and below, this little snake becomes an opalescent purple in color when it prepares to shed its skin. There are no enlarged belly scales and no visible eyes. The tail, which is tipped by a spinelike scale, is nearly as blunt as the head. There is no enlarged anal plate. The Brahminy blind snake is so different from the popular concept of a snake that many folks who see it think it is a strange worm.

Similar species: None in the eastern United States. In the lower Rio Grande valley, one may also encounter the Texas blind snake. This latter species is pinkish to pinkish brown rather than brownish black.

3

File Snakes

File Snakes, family Acrochordidae

One species of this Austral-Asian snake family has been established in Miami, Florida, since the 1980s. Except for two other species of file snakes (neither of which occurs in North America), there is nothing quite like this snake in appearance. At first glance, it resembles a roughened, wrinkled bicycle tire inner tube.

File snakes as a group are nonvenomous. They are strictly aquatic and occur in fresh, brackish, and saltwater habitats. As would be expected, they are awkward on land but graceful when in the water.

The Javan file snake is active at night. Typical night fishing behavior consists of anchoring the tail in the limestone fissures and extending the body, catching fish with a quick sideways snap as the fish passes. If the fish is small, it is quickly swallowed. If the fish is large it is held tightly in a body coil and manipulated for easier swallowing.

The snake may be occasionally seen swimming at night through beds of hydrilla and glasswort. When illuminated by the beam of a flashlight, the snake quickly recoils and almost as quickly, finds subsurface refuge.

The reproductive biology of the Javan file snake in Florida is unknown. It is a live-bearing species; in Asia it has litters of more than 30 babies. It is not known whether it breeds annually or only once every two or three years.

The file snake quest

The drainage canals that crisscross the oolitic limestone substrate of South Florida were dug in an effort to drain the Everglades for dwellings and agriculture. In most cases the canals accomplish their objective reasonably well; too well, in some cases. But even during dry seasons the canals usually hold several feet of water, and they have become sanctuary to water-loving birds, mammals, reptiles, and amphibians. Otters and anhingas forage in the deeper areas. Snail kites search for apple snails and vary their diet occasionally with a small mud turtle. Water snakes of many species prey on green treefrogs and leopard frogs.

And in one small area of one canal, the many species of (mostly introduced) fish provide a repast for a very unusual introduced snake.

This is the file snake, an Australasian native known to be present in Dade
County, Florida, since the 1990s. Although its strictly aquatic habits have
made population assessments impossible, the file snake seems to be quite
rare. It is a snake that we are still attempting to find in Florida.

Because of our lack of success, we thought it possible that the file snake
had been extirpated in Florida, but in 2002 several neonates were brought to
us. They had been found benumbed in the shallows of a South Florida canal
after an unseasonably cold night. There were 7 of them, all about a foot in
length, and all still bearing an umbilical scar that indicated recent birth. Had
it not been for the effects of the unprecedented cold front it is probable that
these neonate file snakes would never have been seen.

But they were, and again our interest has been piqued and our search has
resumed. The finding of those babies has now made the possibility of seeing
one of these amazing fish-eating snakes in the "wilds" of Miami seem a little
more real.

5. Javan File (Elephant-trunk) Snake

Acrochordus javanicus

Abundance/Range: In Florida this is a rare snake. It is known only from Dade
County, Florida.

Habitat: The Javan file snake does not seem to have expanded its range since its

5. Javan File Snake

original introduction. It is still found only in a few manmade ponds (borrow-pits) and canals in Dade County, Florida. This snake's presence in Florida is the result of deliberate introduction.

Size: Female Javan file snakes attain an adult length of more than 72 inches. Males are often only half that size.

Identifying features: The vernacular name of elephant-trunk snake rather accurately describes the appearance of this species. The name of file snake is also eminently accurate, for the scales are rough and filelike. When seen out of water, the file snake imparts the impression of "a size 6 snake in a size 8 skin." The dorsal color is an olive brown with poorly defined olive green mottling, ocelli, or spots. The ventral coloration is lighter. There are no enlarged ventral scales. The rough scales are ideal for penetrating the slime coating and securely restraining a fish when the snake catches and coils around one. The eyes are small and directed dorsolaterally. The nostrils are high on the snout, close together, and directed forward. The tongue has very long forks that separate widely when protruded.

Similar species: None; no other Florida snake is even remotely similar in appearance to the file snake.

4

Boas and Pythons

Boas and Pythons, family Boidae

Although no boa or python is native to the eastern or central United States, two species of small boas occur in the American West.

The world's largest snakes are contained in this family. There are, however, also a great many diminutive species of boas and pythons. All are powerful constrictors.

Boas and pythons have long been staples in the pet trade and are sold annually to hobbyists by the thousands. Three species in particular are very popular: the common boa, the ball python, and the Burmese python.

Some pythons—the various rock pythons, scrub pythons, and anacondas—grow very large. They are interesting and easily kept when babies, but few hobbyists are truly prepared and capable of maintaining them when adult, and it can be extremely difficult to find a new home for them. Zoos generally already have all they need. At the worst, the snake is often released (in violation of state laws) in relatively remote areas, only to cause a furor when days or weeks later it turns up looking in fine condition. Although releases have been reported across North America, the snakes are very cold sensitive so it is doubtful that any survive the winters north of southern Florida or the lower Rio Grande valley of Texas.

Of the three species of exotic boids now known to occur in the central and eastern states, only the Colombian boa is known to have reproduced, and only in Florida. Burmese pythons of all age groups have been found in South Florida, but an incubating female has not yet been found. Discarded/escaped ball pythons are frequently found in apartment complexes and housing developments. Occasional examples of the potentially gigantic reticulated python (adult at up to 30 feet in length!) have been found in Florida. However, there is no indication that this species is established.

The pythons and boas are relatively primitive snakes that retain vestiges of hind limbs. These are visible as a "spur" on each side of the snake's vent. Males have the larger spurs. These are used during the courtship procedures.

These snakes depend on ambush to secure prey. Once secured, the prey is immobilized by the powerful coils until death occurs by heart failure and suffocation.

Hatchlings are opportunistic feeders, eating invertebrate or vertebrate prey. Adults prefer mammals and birds to other prey. In Florida, it is probable that squirrels and opossums are main dietary items of adult boids.

Boas thermoregulate in sunny spots on cool mornings. They are most active in the early evening, but during the truly hot weather may remain active long after darkness has fallen.

The snakes in this family are nonvenomous.

It was a dark night ...

Throughout our lives we have heard tales about huge snakes that stretch from one side of the road to the other. But we always supposed they were just that—tales—or, if accurate, that the roadway involved was inordinately narrow. When I (RDB) was driving through Everglades National Park one evening, I saw in the headlights what looked to be an extraordinarily thick fire hose stretched across the entire pavement. I slowed because I couldn't figure out how a fire hose had gotten out in the 'glades.

As my car slowed, I could see the "hose" was moving. The fore end was now well off the pavement and the rear end was about in midroad. What, I wondered, was I looking at? By the time I stopped the car, grabbed a flashlight, and got out, the object was almost entirely off the road. It was a Burmese python and I estimated its length to be about 14 feet. It was the biggest snake I had ever seen in the wild. What was it doing in Florida? How had it gotten here? What would become of it? Questions for which I then had no answers tumbled forth.

The year of that sighting was 1996; although I had heard that the Burmese python was present in South Florida I had not yet encountered one. Since then I have seen many more, all ranging in size from a little more than 3 feet to about 10 feet.

These increased sightings may be due to the constant release into the wild of unwanted pets. Until a nesting female is found, that will remain the supposition, but there are many among us who now feel that nesting females will be found very soon in South Florida.

Boas, genus *Boa*

Although boas are mainstays of the pet trade, the origin of the population in Dade County, Florida, remains unclear. These relatively large snakes are powerful constrictors that, when adult, should be approached with caution. Although slow moving, they can be formidably defensive if suddenly startled.

The single established species in Florida is a live-bearing snake that, in its natural range, produces large litters (20–50) of relatively large neonates. Its reproductive rate in Florida is unknown.

6. Colombian (Common) Boa Constrictor

Boa constrictor imperator

Disposition: The Colombian boa may be defensive when approached. A large one can deliver a serious bite. If bitten, clean and dress the wounds well.

Abundance/Range: This snake is found with some regularity in Miami-Dade County, Florida, but is still considered an uncommon species. Its range does not seem to have increased in the last two decades. The natural range of this snake is from Mexico to Colombia.

Habitat: The only known population of this snake in the United States occurs in Miami, where it is most frequently seen in pine woodlands among jumbles of limestone boulders.

Size: Most adults seen are in the 60–84-inch range, but this snake has the potential of attaining 144 inches in length. Newborn boas are about 14½ inches in length.

6. Colombian Boa Constrictor

Identification: This is a beautiful and heavy-bodied snake with a pattern that has been likened to an oriental tapestry. Colombian boas are predominantly tan to light brown anteriorly crossed by wide, darker brown saddles that extend as triangles midway down the sides. The saddles widen and shade to a beautiful, variable orange red posteriorly. The red of the tail is separated by white cross bands. The top of the head is the same tan as the body, and bears a darker central longitudinal bar from the snout to the anterior nape. There is a dark bar anterior to the eye extending from the canthus to the lip and a dark eye stripe. The belly is lighter than the back, and is peppered or spotted with darker pigment.

Neonates are paler than the adults, but the colors are the brightest on juvenile to subadult boas. The belly plates are proportionately small.

Similar species: The pattern of dorsal saddles and warm brown body that shades to maroon or red on the tail is distinctive. There are no similarly colored snakes in the eastern or central states.

Typical Pythons, genus *Python*

These powerful constrictors are Old World snakes that are immensely popular in the pet trade. The genus contains seven species, most of them mainstays of the pet trade. The Burmese python and ball python are now found regularly in South Florida and more rarely elsewhere. Their continued presence is due largely to continuing releases or escapes of captives.

The reproductive biology of the pythons is unknown in the wilds of Florida. It is not even known with certainty that any species does breed here.

However, pythons are oviparous. Large species lay large numbers of eggs (Burmese pythons lay 20–60+ eggs the size of goose eggs) while species such as the ball python lay 4–8 eggs. Some species incubate their eggs. They elevate their body temperature (when necessary) a few degrees above that of the surrounding air through periodic muscle contractions.

When frightened or surprised, Burmese pythons can be formidably defensive. Despite their large size, they have been seen to fall prey to alligators.

Although a ball python may hiss when approached, if startled or handled, it usually rolls itself into a ball, head innermost and protected by the coils.

All pythons are nonvenomous.

7. Burmese Python

Python molurus bivittatus

Disposition: Feral examples of this species are often very ready to bite. A large Burmese python can be a dangerous adversary. The constricting coils of a large individual can be life threatening. If bitten, clean the wounds well and sanitize. Large adults should not be approached.

Abundance/Range: Although escapees or released examples have been found in many areas of the country, this is a very cold-sensitive snake. It is likely that only those in subtropical South Florida or the lowermost Rio Grande valley of Texas could be long-term survivors. The natural range of this snake is Southeast Asia, Malaysia, and Indonesia.

Habitat: This snake is most commonly found prowling on summer days in city lots or parks and most show few signs of defensive behavior. Because of their benign behavior, we feel that most of these incidental finds are probably relatively recent escapees or releasees. However, since the 1990s Burmese pythons of all sizes have been found with disquieting regularity in Everglades National Park.

7. Burmese Python

These have, for the most part, been lean and defensively mean, signs that the snakes have been feral for some time. Although this species is not yet known with certainty to breed in Florida, biologists suspect that this might now be happening.

Size: This is the largest snake species found in North America. Most adults are in the 7–12-foot range, but this snake can reach more than 20 feet in length. Hatchlings are about 14–18 inches long.

Identifying features: The normal color of the heavy-bodied Burmese python is a study in browns, tans, and black. Albinos and others with either a fragmented busy pattern or no pattern are popular in the pet trade. The ground color is tan, but the black-edged brown pattern is difficult to describe. There is a pattern of irregular dark dorsal saddles. Lateral blotches of irregular shape are also present and these are often dark edged posteriorly. The comparatively narrow belly scales are light tan. The top of the head is adorned with a dark spearpoint, sharp end at the tip of the snout. There is a dark eyestripe. Neonates are colored and patterned like the adults.

Similar species: Pattern, color, and size should identify this snake. See also the account for the ball python.

8. Ball Python

Python regius

Disposition: This small innocuous python seldom bites.

Abundance/Range: Feral examples are occasionally seen in many cities. This snake is most often reported from Florida. The natural range of the ball python is tropical West Africa.

Habitat: The ball python is most commonly found prowling on summer days in residential areas, suggesting relatively recent escape or release. Whether these little pythons will eventually become established in Florida or the lower Rio Grande valley, or whether this cold-sensitive snake can even survive a normally cold winter, is not yet known.

Size: This is a small but very heavy-bodied snake. Ball pythons are adult at about 60 (rarely to 72) inches in length. Hatchlings are about 10 inches in length.

Identifying features: Like the Burmese python, the much smaller ball python is clad in scales of tan, brown, and black (aberrant colors and patterns have been developed). The ball python has an irregular dorsal pattern but a regular lateral pattern of large, *rounded*, light markings. Most light lateral blotches contain one or two small dark ovals. The top of the ball python's head is dark. The belly scales are narrow and light tan. There is a dark eyestripe. Neonates are colored and patterned like the adults.

8. Ball Python

Similar species: This snake may be confused with a small Burmese python. Please check account number 7 for comments on the Burmese python. Burmese pythons usually have regular, well-defined dorsal blotches and triangular or rhomboidal lateral spots with dark posterior edging. The light lateral spots of a ball python are more rounded and bordered on all sides by black pigment. Also, the top of a ball python's head is dark and does not bear a spearpoint marking.

5

Colubrid Snakes

Typical or Colubrid Snakes, family Colubridae

As currently defined, the colubrids are a large and unwieldy assemblage to which most snakes are assigned. The taxonomic assignments are rendered somewhat more workable by placing allied genera into subfamilies. In this book the colubrids are placed within five subfamilies.

Although the colubrids are generally considered harmless, the members of several subfamilies have toxic saliva and some have enlarged teeth at the rear of the upper jaw. All should be handled carefully. Among eastern

Figure 5.01. Eastern Hog-nosed Snake. Illustration by K. P. Wray III.

and central species known to have some degree of toxicity are the various hog-nosed snakes, some garter snakes, and some water snakes.

Other species such as the crowned snakes, lyre snakes, and ring-necked snakes are known also to produce toxins, but in North America these species are too small to be of concern to humans.

The colubrid snakes vary widely in appearance and lifestyles. Many are short, stocky, and terrestrial. Some are glossy-scaled burrowers. Others are slender speedsters that may as readily ascend trees as seek refuge on the ground.

Racers and Related Snakes, subfamily Colubrinae

The Colubrinae include a number of snakes that are of diverse appearance and habits. In eastern and central North America, snakes of ten genera are assigned to this group. Among these are some of the fastest snakes found in the United

States—the racers and the whipsnakes. Additionally, there is the threatened eastern indigo snake, a large racer look-alike that traditionally travels over a range encompassing almost 500 acres, a range so large that it is almost impossible to offer the species full protection from vehicles and nervous humans. The little green snakes are insectivorous fencerow species that are so well camouflaged when among greenery that unless moving they are almost invisible. Some, such as the crowned snakes and ground snakes, are tiny burrowers that feed principally on centipedes or insect larvae. All members of this group in the central and eastern states are egg layers.

Racers, genus *Coluber*

The various subspecies of the black racer are found from the Atlantic to the Pacific coasts, but are absent from vast tracts in our southwestern and north central regions.

Racers are generally regarded as terrestrial serpents, but climb readily and well.

The majority of the ten subspecies of the black racer occur in the eastern and central United States (only one race is restricted to our west). All are noted for their active and alert demeanor, readiness to bite if molested, and considerable speed. These snakes may even occasionally stand their ground or approach a person (especially if they feel cornered). They do so with head, neck, and anterior body held in an S shape well above the ground, often feinting, striking, and vibrating the tail in dried vegetation as they approach. This can be disconcerting to those who understand snakes, and terrifying to those who don't.

The alacrity with which a startled black racer disappears into cover can be nearly as disconcerting as its occasional aggressively defensive stance.

All are diurnal but often remain in hiding unless the sun is shining brightly. Although most of them are a unicolored black, gray, blue, or brownish green dorsally when adult, two races have the dark ground color interestingly and variably patterned with spots and blotches of light pigment. The hatchlings and juveniles of most are very strongly patterned with dark dorsal blotches (or crossbars) on a tan to light brown or grayish ground color.

Hardly could a scientific name be more erroneous than the specific designation *constrictor* that has been bestowed upon this species, for the racers are, most emphatically, *not* constrictors. They merely grasp their prey—usually frogs, lizards, smaller snakes, nestling rodents and birds, and even insects—

and swallow it alive, occasionally making a cursory attempt to immobilize a struggling prey item with a loop of their body. But they never constrict. When hunting, racers often keep their head elevated above grass levels (periscoping) searching visually for their prey.

All members of this genus are oviparous, and produce about 5–20 eggs with a characteristic "pebbly" shell. The secluded nesting site (occasionally used by multiple females) is beneath or in moisture-retaining logs or debris, natural or manmade. The incubation duration varies from about 48 to 65 days, dependent on warmth and moisture. The scales are smooth, arranged in 17 rows at mid-body (usually dropping to 15 rows anterior to the vent), and the anal plate is divided.

Expect to see these snakes in and near open woodlands (especially near clearings or at woodland edges), in pastures and meadows, at bog edges, along roadsides, and in sandy lands. They are often seen in suburban and rural yards.

All are nonvenomous.

Remembrances of black racers and indigo snakes

One of the first snakes I remember from childhood was a northern black racer (then known to me simply as a black snake) that darted in and out of the tall vegetation in the neighborhood churchyard. I managed to get between the big snake and its cover one day and was rewarded by seeing the snake assume a defensive S and come straight toward me. It was a memorable occasion for a child, but one that has been repeated many times since by racers and whipsnakes of several species in my presence. These snakes simply do not react favorably to unsolicited overtures.

Their bigger relative, the eastern indigo snake, is far less flighty and a whole lot more tolerant of intrusions. In fact, in many cases indigos could actually be called "laid back." In the 1950s and 1960s, when I began traveling to Florida during school vacations in winter, indigos were a common species. But then, in those days, the open land that the indigos needed for survival was also plentiful. Nonmechanized orange groves with open irrigation canals replete with frogs and water snakes were a favored habitat of indigo snakes. One large grove down in Davie, Florida, was prime habitat, and on any warm, sunny winter day we could expect to see at least one, and sometimes several, indigos foraging along the shallow canals.

Those were indeed the "good old days …"

9. Northern Black Racer

Coluber constrictor constrictor

Disposition: This feisty snake will bite readily if cornered or molested and will sometimes even approach an antagonist. A bite from even a moderately large individual can draw blood, but the wound produced is hardly more than a scratch.

Abundance/Range: This is one of the most abundant and successful snakes in eastern North America. This ubiquitous snake ranges southward from southern Maine and central Wisconsin to North Carolina and northeastern Alabama.

Habitat: The northern black racer is a wide-ranging habitat generalist. It may be encountered in suburban fencerows, rural gardens, fields, meadows and pastures, and open woodlands. It may be found on high ground or near pond and swamp edges. It is usually seen on the ground but readily ascends shrubs and shrubby trees, at times climbing high above the ground. It shelters beneath natural and manmade debris when preparing to shed and to thermoregulate.

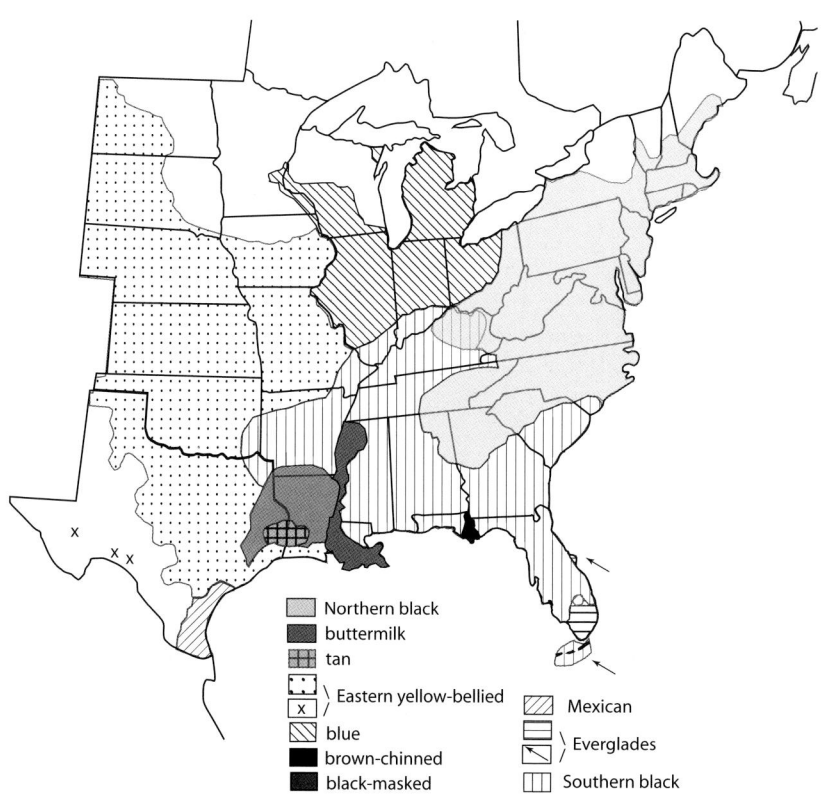

Northern black
buttermilk
tan
Eastern yellow-bellied
x
blue
brown-chinned
black-masked

Mexican
Everglades
Southern black

9. Northern Black Racer

Size: The northern black racer is adult at 3–4 feet, may attain 5 feet, and rarely attains 6 feet. The very slender hatchlings vary from 10 to 12 inches in length.

Identifying features: Except for a variable amount of white on the chin and brown(ish) irises, adults of this snake are a lustrous, satiny (as opposed to shiny) black both above and below.

Hatchlings are very different than the adults in appearance. Young are patterned with a series of reddish to brown dorsal blotches against a gray ground color. The dark blotches are best defined anteriorly. Small dark spots occur on the sides and on the grayish venter. With growth, the ground color quickly darkens and the blotches are overwhelmed and obscured.

Similar snakes: The southern black racer (account 18) is of very similar appearance and best identified by range. Other racers are blue or spotted, or have differentiating facial markings. The more robust eastern indigo snake has a more southerly distribution and has shiny, polished scales. Black-phase eastern hognosed snakes have a sharpened, keeled, and upturned rostral scale. Baby rat snakes have small eyes, a checkered belly, and relatively large dorsal blotches.

Additional subspecies

10. The beautiful Buttermilk Racer, *Coluber constrictor anthicus*, is variably patterned. The body color of this subspecies ranges from just on the blue side of black to black. In the center of its range, where the influence of the tan racer may be seen, the body color may be olive. The flecks, which are variable in

10. Buttermilk Racer

number and haphazard in pattern, are olive, light blue, or occasionally, yellowish. This subspecies is found eastward from eastern Texas to central Louisiana and southern Arkansas.

11. Tan Racer. Photo by Tony Hibbitts

11. The Tan Racer, *Coluber constrictor etheridgei*, is very similar in appearance to the buttermilk racer, but has a tan to light brown ground color and a highly variable number of somewhat lighter spots. Some specimens are largely without light flecking and others are very prominently marked. This snake of southeastern Texas and adjacent Louisiana is often of very drab appearance. It is firmly associated with stands of longleaf pine.

12a. Eastern Yellow-bellied Racer

12b. Eastern Yellow-bellied Racer, juvenile

12. The Eastern Yellow-bellied Racer, *Coluber constrictor flaviventris*, is a very pretty, nonflecked snake with a grayish blue, brown, or greenish back and a yellow to off-white belly. It has an immense but spotty range, being found from southern Louisiana to central Texas, then northward to central Montana, central North Dakota, and extreme southern Saskatchewan.

13. Blue Racer

13. The Blue Racer, *Coluber constrictor foxii*, ranges southward and eastward from southeastern Wisconsin to southern Illinois and northeastern Ohio. It can be very much like a black racer in coloration in the eastern portions of its range or like a yellow-bellied racer in the west, but is often a very decided deep blue color dorsally and a lighter blue on the sides and belly. The chin (and often the throat) is white.

14. Brown-chinned Racer

14. The Brown-chinned Racer, *Coluber constrictor helvigularis*, has a very limited range in the Apalachicola River valley of Florida's panhandle. It is black both dorsally and ventrally and the chin and throat are pale brown to variably brown and white. The brown may also be present on the lower half of the upper labial scales. The hatchlings have a ground color of gray and bear large, dark-edged, reddish dorsal markings. There are two rows of small reddish markings along each side.

15. Black-masked Racer

15. The Black-masked Racer, *Coluber constrictor latrunculus*, is restricted in distribution to a narrow north-south swath of western Alabama and Louisiana. The back is blackish to dark bluish gray and the belly is usually a lighter bluish gray. This snake derives its common name from the (usually) prominent mask of darker color that runs through the eye.

16. Mexican Racer

16. Much about the Mexican Racer, *Coluber constrictor oaxaca*, is different than the other subspecies of the black racer. The Mexican racer is a blue-gray to greenish brown dwarf form. It is adult at 2–3 feet in length. The belly may be pale yellow, off-white, or yellowish green. The greenish gray hatchlings are patterned with narrow dark cross bands (not light centered blotches) anteriorly. These break into dark spots posteriorly. This very alert and active little snake occurs along the coastline of southeastern Texas and in the Big Bend region of Texas. In the latter, a very arid area, this racer is restricted to the vicinity of waterholes, stock tanks, and river edges.

17. Everglades Racer

17. The Everglades Racer, *Coluber constrictor paludicola*, is a light-colored snake. The dorsal ground coloration is variable, usually a grayish green to a grayish brown, but never jet black. The lower half of the supralabials, the lower labials, the chin, and the venter are bluish white to bluish gray. This color sometimes carries over to the lowest row of lateral scales. The hatchlings have a ground color of gray to grayish tan and are adorned with many dark dorsal saddles. The saddles are very pronounced anteriorly and indistinct to absent posteriorly. The sides are liberally flecked with dark pigment. The venter is lighter. The eyes are proportionately large. It is restricted in distribution to Everglades habitat (past and present) and a population of racers of similar appearance also occurs near Cape Canaveral.

18. Southern Black Racer

18. The Southern Black Racer, *Coluber constrictor priapus*, is almost identical in appearance to the northern black racer. The southern snake may have a little more white on the chin and throat (especially in Florida) than its northern relative, but the truly differentiating features—the morphology of the hemi-

penes—are internal. Use range as your primary identification tool. This ubiquitous snake ranges westward from southeastern North Carolina and southern Florida (including the Keys) to southeastern Missouri and northeastern Texas.

Whipsnakes, genus *Masticophis*

Whipsnakes are closely allied to the racers of the genus *Coluber*. Whipsnakes are alert, slender, very speedy serpents that often seek prey by periscoping their head above the surrounding grasses and ground plants. Their vision is acute, and the whipsnakes seem to rely as much on visual as on chemical cues when pursuing prey. All manner of prey—from frogs, lizards, and smaller snakes (including those of their own kind), to baby turtles and tortoises, small mammals and birds—are opportunistically eaten. Because they are alert to visual cues, whipsnakes (some are called coachwhips) readily locate and predate the nests of birds in shrubs and low trees. The food is not constricted.

There are only three species of whipsnakes in the eastern and central states and they are of rather similar appearance. The scale row count varies from 15 to 17 at midbody. The scales are smooth, the anal plate is divided. While rather narrow (bluntly lance shaped), the head is deep and well defined from the slim neck.

Although they would prefer a rapid and uneventful escape to a standoff, whipsnakes vibrate their tail, assume a striking S shape, and bite strongly if cornered. Some may even approach an offending object (such as a human) as they strike. This can be a daunting experience to someone unfamiliar with the ways of snakes. However, if grasped and restrained, these snakes may actually play dead, making their body rigid, holding the head at an unnatural angle, and remaining stationary even when placed on the ground.

These snakes may ascend shrubs or small trees to escape.

All of the whipsnakes (and racers) lay eggs with a characteristic pebbly shell. A clutch consists of 5–12 eggs. The eggs are secluded beneath or in moisture-retaining logs or debris of either natural or manmade origin, or may be laid in an unused mammal or reptile burrow. The incubation duration varies from about 48 to 65 days, dependent on warmth and moisture.

Look for these sand and scrub speedsters in the proximity of rangeland, old citrus groves, scrub oak ridges, dumps, thorn scrub, and other similar areas.

All whipsnakes are nonvenomous.

19. Eastern Coachwhip Snake

Masticophis flagellum flagellum

Disposition: This is a feisty snake that will bite readily if cornered or molested and will sometimes even approach an antagonist.

Size: This is a very slender snake often judged to be smaller than its actual length. Although it may occasionally attain a length of 8½ feet, it is more usually 5–6½ feet long. The very slender hatchlings are usually 13–16 inches in length.

Abundance/Range: Although still common in some areas, the eastern coachwhip is not seen as often as in bygone years. This snake ranges eastward from eastern Texas and southeastern Kansas to central eastern North Carolina and Florida. There is apparently a range hiatus along the floodplain of the Mississippi River.

Habitat: This is a snake of sandy rangelands, sandhills, and scrublands. Look for it along overgrown fencerows, old fields, sandy pastures, open woodlands, sprawling prairies, and old citrus groves. It may even be found in sandy areas within sight of the ocean. The coachwhip is usually seen on the ground but readily ascends shrubs and shrubby trees, at times climbing high above the

eastern
lined
western

19. Eastern Coachwhip Snake

ground. It shelters beneath natural and manmade debris or in mammal or gopher tortoise burrows when preparing to shed and to thermoregulate.

Identifying features: Age-related color and pattern changes are notable. Hatchling eastern coachwhips have a strongly light and dark patterned face, and dark dorsal cross bands that are separated by about 2 rows of light scales. These markings fade with maturity. There are several adult color phases, but on all, the head is the darkest and the tail the lightest. One phase is a solid sand tan (sometimes the top of the head is dark). A second is black anteriorly, fading to tan about a third of the way back. This phase often has the light scales outlined with dark, a characteristic that produces the braided look of an old coachwhip and has given the snake its common name. A third phase with a solid black body and a (usually) dull reddish tail is also known. The belly may be tan, grayish tan, or very dark and is lightest below the tail. Some striping may be visible on the throat. The eyes are large, and widened supraocular scales shade the eyes and give the snake a sullen look. The eye is usually yellow or reddish. The scales are smooth, in 17 rows (13 rows just anterior to the vent—an important factor when trying to decide whether you have a black coachwhip or a black racer in hand), and the anal plate is divided.

Similar snakes: The western coachwhip can be confusingly similar to its eastern relative (see account 24). Juvenile western coachwhips usually have the dark crossbars separated by 3 rows of light scales. Use range as an identification aid. The southern black racer has 15 scale rows just anterior to the vent. Baby racers have large dorsal saddles. The dorsal markings of the coachwhip are narrower. Baby rat snakes have small eyes, a checkered belly, and large dorsal blotches.

20. Lined Coachwhip Snake

Additional subspecies

20. If the Lined Coachwhip Snake, *Masticophis flagellum lineatulus*, occurs in our area at all, it is only in the vicinity of El Paso, Texas. It does occur in New Mexico and parallels the Mexican side of the Rio Grande (the Rio Bravo in Mexico) along much of the West Texas border. From there it ranges southward far into Mexico. On this coachwhip race the center of each anterior dorsal and lateral scale bears a narrow longitudinal black mark. There is often quite considerable dark pigment outlining posterior scales. The subcaudal scales may be tan with a pink suffusion or a rather rich pink or salmon.

21. Somewhat smaller than the eastern coachwhip, the Western Coachwhip Snake, *Masticophis flagellum testaceus*, attains an adult size of 6½ feet. It is both of common occurrence and of variable coloration. This snake ranges southward from southwestern Nebraska, through the western three-quarters of Texas, to interior Mexico. The western coachwhip is most often straw tan (with or without narrow darker bars) in the eastern portion of its range, but in the west it may be pinkish, orange red, or occasionally banded pink and tan, again with or without narrow darker bands. Narrow darker scale edges may be present on any color phase. This subspecies is less apt to shade to different colors than the eastern coachwhip. The juveniles of the western coachwhip are usually paler than the adults and have dark crossbars separated by about 3 rows of light scales. The nose of all sizes and ages is often quite dark. Two rows of small dark spots

21. Western Coachwhip Snakes

extend rearward from the throat to the anterior belly. Use the range map to help you identify this snake.

22. Schott's Whipsnake

Masticophis schotti schotti

Disposition: Like all racers and whipsnakes, this snake may try to bite if carelessly restrained.

Abundance/Range: Although not commonly seen, Schott's whipsnake does not seem particularly rare. Well camouflaged and wary, this snake more often than not glides into the safety of the thorn scrub before it is seen by a casual observer. This snake occurs in south Texas (south of the Balcones Escarpment), crossing the Rio Grande to north central Mexico.

Habitat: This is a snake of aridland habitats, but it may sometimes be found

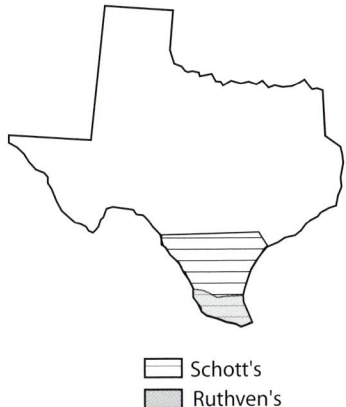

□ Schott's
■ Ruthven's

22. Schott's Whipsnake

near stock watering tanks and waterholes. It is a resident of thorn brush habitats, rocky and densely overgrown hillsides, and similar terrain that is difficult to negotiate. Roadside dumping sites, especially those with old furniture, appliances, and carpets, are also favored habitats. The easily accessed areas of seclusion are ideal when the snake is preparing to shed its skin (ecdysis) or must thermoregulate.

Size: It is the 3½–4½-foot examples of this snake that are most commonly seen, but Schott's whipsnake may occasionally attain 5½ feet in length. Hatchlings are about 12½ inches long.

Identifying features: Although dark, dorsally and laterally this snake is decidedly on the gray or olive gray side of black. Several rows of dorsal scales are edged posteriorly with white. This feature is not easily discernible unless the snake is in hand and the scales can be manipulated. A single, prominent, white dorsolateral stripe is present on each side and another less conspicuous one parallels each edge of the belly scales. The upper labials (lip scales) and chin are white. This shades to off-white on the anterior belly, then to light gray on the posterior belly. The underside of the tail is pink to salmon. The sides of the face are flushed with pinkish red; a variable amount of pinkish red is present on the neck and anterior side and may be most intense along the edge of the ventral plates. Juveniles are paler examples of the adults.

Similar species: Use range as an identification aid. This is the only prominently striped whipsnake in south Texas. Except when it is young, Ruthven's whipsnake lacks well-developed striping. The Central Texas whipsnake is black and has dorsolateral striping of variable definition and irregular width. The desert striped whipsnake is black with 4 white stripes and enters our area only near El Paso, Texas.

Comments: Whipsnakes are slender and, with their high metabolisms, designed for immediacy in their responses. They are not easily approached.

Schott's whipsnake seems able to tolerate very hot temperatures and may actively forage until noontime or after. During the hottest times on the hottest days the snake may stay in the shadows cast by dense brush or seek the comparative coolness of an available burrow.

23. Ruthven's Whipsnake

Additional subspecies

23. Ruthven's Whipsnake, *Masticophis schotti ruthveni*, is the less colorful representative of this species from extreme south Texas. It also occurs in northeastern Mexico. If stripes are present on adults they are indistinct (especially posteriorly) against the blue-gray to army (or darker) green ground color. The scales of the dorsum are often edged with quite a bright yellow. The belly is yel-

low anteriorly, grayish at midbody, and pinkish posteriorly. The scales beneath the tail are bright orange red. The lips are white to pale yellow. Juveniles are quite prominently striped.

24. Desert Striped Whipsnake

Masticophis taeniatus taeniatus

Disposition: This feisty snake will bite if restrained.

Abundance/Range: This whipsnake, common to the northwest of the range covered by this book, enters our area only in extreme western Texas.

Habitat: This is a species of arid, scrubby, montane fastnesses as well as sea level chaparral. It occurs to elevations of at least 9,400 feet. Although found in semi-arid and aridland regions, this snake is often associated with permanent and intermittent streams and marsh edges and in the vicinity of stock tanks.

Size: Although usually smaller, the desert striped racer has been known to attain a length of 6 feet. Hatchlings are a foot or more in length.

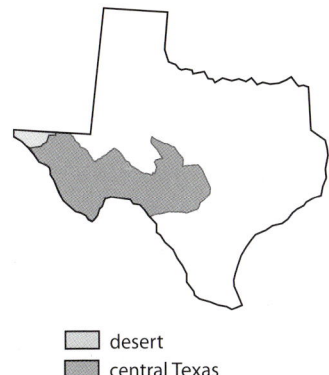

■ desert
■ central Texas

24. Desert Striped Whipsnake

Identifying features: Throughout much of its range the ground color of this pretty, agile snake is black. Occasionally the ground color is of some shade of brown or gray. The 4 thin stripes are very well defined. Close examination of the upper (dorsolateral) stripe will disclose a very thin black line (or dashes) running lengthwise through the center, actually dividing it into 2 very narrow stripes. The ventrolateral stripe is less conspicuous than the upper one, but still usually very visible, and separated from the yellowish ventral scales by a thin black line. The larger facial scales are edged with white. The labial (lip) scales are mostly white, some occasionally bearing a tiny black spot. The belly is an off-white anteriorly and yellowish gray (often with darker blotches) from midbody to vent. The subcaudal scales (those on the underside of the tail) are orange red. Juvenile striped whipsnakes are very strongly white striped against a greenish ground color. The belly is predominantly pale yellow.

Similar species: Schott's whipsnake is also a 4-striped species, but the dorsolateral stripes are not divided lengthwise by dark pigment and it has considerable red on the sides of the face and neck. The ranges of the two do not overlap. The various patch-nosed snakes have a tan to buff ground coloration and prominent dark stripes.

Additional subspecies

25. Unlike the desert striped whipsnake, the Central Texas Whipsnake, *Masticophis taeniatus girardi*, is widely distributed in western and central Texas, including the Trans-Pecos region and the Edwards Plateau. It is also of wide distribution in northern Mexico. The pattern borne by this snake is difficult to describe. We will simply say that it is similar in color to the desert striped whipsnake, has both stripes and patches of white, and may have poorly defined light cross bands. The overall pattern is less intense and more irregular than that of the desert striped whipsnake. Use the range map as an identification aid.

25. Central Texas Whipsnake

Indigo Snakes, genus *Drymarchon*

This genus is currently in taxonomic disarray. Some researchers regard what many believe to be subspecies as full species. At the moment we decline this speciation that is delineated by range and continue to apply trinomial designations.

Indigo snakes are among the largest of colubrines, and may occasionally exceed a length of 8 feet. They are nonconstrictors, overpowering their prey with their strong jaws and, if necessary, immobilizing the prey beneath a loop of the body. Prey includes other snakes, lizards, baby turtles, amphibians, small mammals, and ground-dwelling birds.

Indigo snakes are active in autumn, winter (in some areas), and spring, but seldom seen during the heat of midsummer.

The males of reproductively active indigo snakes fight savagely and may inflict severe wounds on each other. Clutches consist of 3–12 eggs. Like a racer's eggs (but much larger), those of the indigo have pebbly shells. Incubation takes about 2 months.

The eastern indigo snake is an active snake that utilizes an immense range. Thanks to advanced tracking methods, much about the home range and territoriality of the indigo snake has been learned and confirmed. Other aspects of its home life remain enigmatic. The range of adult indigo snakes varies with the weather. During the warm weather, adult males are known to utilize ranges of about 400 acres each! Many territories are much larger. During cold weather the snakes are comparative homebodies, occupying ranges of 50 or fewer acres. This need to wander widely almost invariably brings these snakes into contact with humans and vehicles. All too often the snakes are injured or killed during these encounters.

Indigo snakes occupy the burrows of armadillos and gopher tortoises, stump-holes, or almost any other safe haven when resting. They seem to know their territories intricately, if possible heading straight to a secure area when threatened. Indigos respond in alert manner to most visual stimuli, be it as small as the movement of a frog or rat, or as overt as the approach of a human.

Although most indigo snakes will attempt to escape if approached by a predator (including a human), some, if truly surprised, will stand their ground. With head, neck, and anterior body held in an S well above the ground, a frightened indigo may strike repeatedly and vibrate its tail. If the opportunity presents itself, the snake will turn and disappear quickly into surrounding cover.

These snakes eat frogs, lizards, smaller snakes, rodents, and birds. When hunting, eastern indigos often elevate their head above grass levels (periscoping) and search visually for their prey.

Look for both races of the indigo snake in scrublands, sandy open woodlands, coastal dune areas, unmanicured citrus groves, agricultural areas, and strips of land along irrigation and drainage canals.

This is a nonvenomous snake.

26. Eastern Indigo Snake

Drymarchon corais couperi

Disposition: Indigo snakes usually have a mellow disposition. If very frightened they may occasionally assume a striking S, but most individuals will allow themselves to be handled without displaying any defensive behavior. However, the indigo snake does have powerful jaws, and a bite from a large specimen will assuredly draw blood.

Size: This is the largest snake, not only of the eastern and southern states, but also of North America. An average adult is 5½–6½ feet in length and the record size is 8 feet 7½ inches! Hatchlings may be 1½–2 feet long.

Abundance: Once a relatively common species, the eastern indigo snake is now seldom seen (although it does remain rather common in suitable pockets of habitat) and is considered a federally threatened species. It is most common

eastern
Texas

26. Eastern Indigo Snake

in Florida and uncommon to rare in adjacent southern Alabama and eastern Georgia.

Habitat: This is a wide-ranging and adaptable snake. It may be encountered in fields, meadows, pastures, open woodlands, and agricultural areas. It is also found in old citrus groves or along canals, habitats that often harbor populations of the amphibians, reptiles, and small mammals upon which this snake feeds. This is a terrestrially oriented snake. It shelters beneath natural and man-made debris, or in burrows or other such lairs, when preparing to shed.

Identifying features: This snake looks like a shiny, heavy-bodied racer. Although the belly color may vary, the dorsal color of the adult eastern indigo is an overall shiny black with indigo overtones. The snout and cheeks may be mostly black or strongly orange red. The belly of the adult indigo may be mostly black (with a red-orange to whitish chin) to as much as 50 percent red (anteriorly). Age-related (ontogenetic) color and pattern changes occur. Hatchlings may have an almost all orange head, be quite blue in ground color, and have prominent light banding or be almost solid black. With growth those that have brighter colors fade until by the time the snake is 2½ feet in length it has assumed its adult coloration.

Females have smooth, shiny scales, but adult males have some of the dorsal scales weakly keeled. Scales are in 17 rows, and the anal plate is undivided.

Similar snakes: Black phase eastern hog-nosed snakes have a sharpened and vaguely upturned rostral scale and dull scales. Black racers have a white chin and a satiny (not shiny) finish on their scales. The speckled racer has highlights of turquoise and yellow and is found in the United States only in the lower Rio Grande valley.

27. Texas Indigo Snake

Additional subspecies

27. The Texas Indigo Snake, *Drymarchon corais erebennus*, looks like a faded version of its eastern relative. The body coloration ranges from a hazy blue black to black with olive overtones. A pattern of crossbars is visible anteriorly. Black lines radiate downward from the eye along the labial scale sutures. The belly is largely olive brown, often with a rose or orange patina on the chin and throat. This state-protected snake is found in southern Texas and northeastern Mexico.

Speckled Racer, genus *Drymobius*

Only a single species of this Neotropical snake genus occurs north of the Mexican border, where it is found in extreme southern Texas. The speckled racer is of moderate size, and very like a typical racer in demeanor. It hunts both visually and by scent, preferring frogs and lizards (with the emphasis on frogs) as prey. Nestling rodents and nestling ground-nesting birds are also eaten. Like a racer, it periscopes its head above the vegetation when danger is sensed or when foraging. If cornered this snake may coil and vibrate its tail, producing a whirring sound.

A single clutch of up to 10 eggs is laid annually. Incubation takes 55–65 days. The nesting site is a moisture-retaining pile of plant debris, a depression beneath a log, or other such well-concealed spot.

This snake is wary, agile, and alert. It usually sees an approaching human and glides to safety long before it is seen. We have found these snakes beneath piles of fallen palm fronds as well as amid exotic ground cover plants. They may bask in the sun on trails or openings in the morning, but seem ever alert. They are most common where there are permanent water sources, and may even drink at ground-level birdbaths in suburban and rural yards.

This is a nonvenomous snake.

28. Northern Speckled Racer

Drymobius margaritiferus margaritiferus

Disposition: This snake will bite if carelessly restrained.

Abundance/Range: In the United States this uncommon racer occurs only along the lower Rio Grande, ranging from there southward to northern South America. It is protected by the state of Texas.

Size: This fast and alert snake is adult at 3 feet in length, but occasional specimens may reach 4 feet. Hatchlings are about 9 inches long.

Habitat: In Texas, this is a species of hot and humid habitats. It is usually associated with the few patches of remaining gallery forest, but has also been seen in riparian scrub and in meadows. In Latin America it is more of a habitat generalist.

Identifying features: This is a fast, alert, and slender snake. At first glance, it may appear to be simply black in color. However, given a good look the speckled rac-

28. Northern Speckled Racer

er will be seen to have prominent turquoise and yellow highlights that will be displayed differently depending on which way the snake is curving. There is a spot of yellow on each scale and the base of each scale is blue. The scales on the back are weakly keeled; those on the sides are smooth. The top of the head is unspotted and shades gently from olive on the snout to bluish green on the rear of the head. A wide black bar is present behind each eye. The belly is white or yellow (rarely the trailing edge of each scale will be black). The subcaudal scales (beneath the tail) are edged with black posteriorly. The anal plate is divided. Hatchlings are quite like the adults in appearance, but paler.

Similar species: None; there are no similarly colored snakes in the United States.

Green Snakes, genus *Opheodrys*

The green snakes are the brightest green of any American serpents. The venter of both species is white to yellowish or of the palest whitish green. The rough green snake is of southeastern distribution and the smooth green snake is the northern representative of the genus. The rough green snake is a slender, usually slow-moving, arboreal species that feeds on caterpillars, crickets, and spiders. Although often associated with tall grasses, shrubs, and low trees, rough green snakes have been found high in the canopy. Field-edge and roadside trees and shrubs are favored habitats.

While having a similar diet, the smooth green snake is more terrestrial, often seeking its food among the grasses and rock jumbles in overgrown fields. Like a racer, the rough green snake often periscopes while visually searching for its prey. The smooth green snake does not seem as apt to do so. If approached, these snakes stop moving and depend on their color for camouflage. If grasped, both species of green snake will indulge in a mouth-gaping defensive display but almost never bite.

Because of its periscoping practices, the vision of the rough green snake is thought to be acute. The scales of the rough green snake are keeled and in 17 rows; those of the smooth green snake are unkeeled and in 15 rows. The anal plates of both species are divided.

Both species are oviparous. A normal clutch consists of 5–12 eggs, secluded beneath or in moisture-retaining vegetation—alive or dead—or in crevices and splits in the trunks of living trees. The eggs have a rough but not pebbly shell.

The green snakes are nonvenomous.

29. Northern Rough Green Snake

Opheodrys aestivus aestivus

Disposition: This snake is quiet and inoffensive.

Abundance/Range: Although still a common species, the rough green snake is now less regularly seen than it was in past years. The rough green snake ranges over much of the eastern United States, from central New Jersey to eastern Kansas and eastern Texas to central peninsular Florida.

Habitat: This snake may be encountered in suburban fencerows and gardens, old fields, and woodland edges, and along vine-covered railroad beds. The rough green snake is usually seen in shrubs and among vines. When preparing to shed its skin, this snake may take shelter beneath natural or manmade debris.

Size: A very slender snake, the northern rough green snake typically attains a length of 2–2¾ feet. Occasional specimens attain a length of 3¼ feet. The hatchlings usually measure about 6 inches long.

Identifying features: Age-related color changes in this racer relative are obvious, but occur so soon after hatching that they are seldom noticed. Hatchlings are bluish green to dull green, but by the time the postnatal skin shedding has occurred (only a few days after hatching) these snakes have already assumed

X Northern
Florida

29. Northern Rough Green Snake

their leaf green coloration. Interestingly, dead specimens quickly lose their green coloration and take on an overall grayish appearance.

As a species this is one of the most readily identified snakes in the Southeast (identifying subspecies is more difficult). The upper lip is edged with white or yellow. The yellow eye is large. The keeled dorsal and lateral scales, leaf green coloration, and immaculate white to yellow belly are diagnostic. The scales are in 17 rows, and the anal plate is divided.

The characteristics that separate the northern form from its South Florida relative are minimal and impossible to ascertain unless the snake is in hand. The primary difference is found on one scale on the side. Counting up from the 7th belly scale from the front, the scale in the 3rd row is usually smooth on the northern race, whereas the same scale is usually keeled in the Florida race.

Similar snakes: Only the smooth green snake assumes an equally uniformly green hue. However, as indicated by its common name, it has smooth, non-keeled scales on the back and sides.

Additional subspecies

30. The Florida Rough Green Snake, *Opheodrys aestivus carinatus*, is a poorly differentiated race. It occurs on the southern peninsula from the latitude of Hillsborough and Brevard counties to Key West in Monroe County. Use range as a primary tool when trying to identify this snake to subspecies. The presence or absence of a keel on a single scale on each side is the criterion. Usually, the scale in the 3rd row of body scales above the 7th belly scale from the front is keeled on the Florida race and not keeled on the northern race. The feature is

30. Florida Rough Green Snake

impossible to ascertain unless the snake is in hand, and even then will often require the aid of a magnifying glass.

31. Smooth Green Snake

Opheodrys vernalis

Disposition: This snake almost never bites.

Abundance/Range: This beautiful little snake seems common to abundant is some areas and uncommon to rare in others. It has more disjunct populations than any other snake in eastern North America. Rather than trying to designate these, we ask that you check the range map.

Habitat: This snake may be found from sea level to quite considerable elevations. Look for this little insectivore in open rock-studded fields, in fencerows, at the edges of boreal bogs, along woodland edges, and on the banks above brooks. The single most important habitat component is natural or human-generated ground-surface hiding places that retain just a bit of moisture.

Size: This tiny, slender snake is adult at 12–18 inches in length, but like all

31. Smooth Green Snake

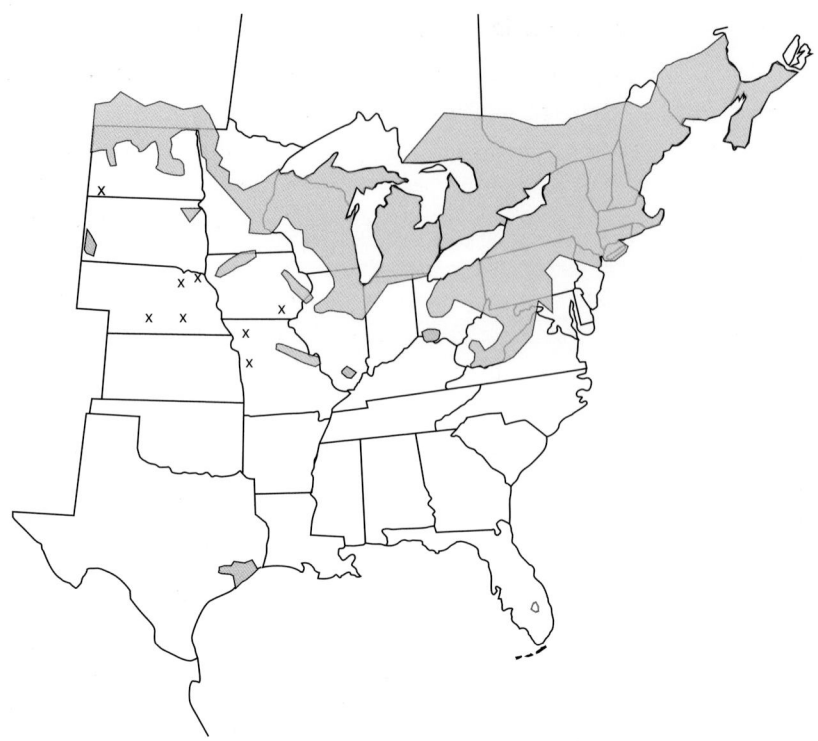

snakes it continues growing throughout its lifetime. Because of this, an occasional example may attain 2 feet in length; the record size is 26 inches. Hatchlings vary in size. Some may be as small as 4 inches in length at hatching, and others may be 6½ inches long.

Identifying features: This smooth-scaled snake is leaf green (rarely tan or light brown) above and on the sides and pure white to pale yellow below. The lower portions of the upper lip scales and the lower lip scales are yellowish. Hatchlings are bluish to olive brown. Those about to shed their skin are quite dull in color. Dead examples also quickly lose their green coloration, becoming gray to bluish in color.

Comments: Much about the reproductive biology of this snake is unusual. Females retain the developing eggs long into the incubation period. Because of this, hatching may occur in as few as 4 days, or as many as 35. Communal nesting has been well documented.

Similar species: The rough green snake, the only other American snake that is a brilliant leaf green in color, has keeled body scales.

Patch-nosed Snakes, genus *Salvadora*

Two species in this genus of diurnally active racer relatives occur in central North America. Both range from Texas westward.

Both are basically sand-colored serpents that seldom exceed a yard in length. A linear pattern of dark and light stripes is the norm. Because of variable coloration, identification as to species can be difficult. Use range as the initial identification tool. The chin scales of the two species are arranged differently, but determining that arrangement will, of course, require having the snake in hand. However, catching a patch-nose can be daunting, considering the speed and agility with which they move through and between the spine-armed plants of their arid homes.

Foraging patch-nosed snakes are very motion oriented. One will ignore a nearby nonmoving lizard but will immediately pursue a moving one. Fast-moving lizards such as whiptails and spiny lizards are typical prey. Small snakes are also eaten, as are reptile eggs. It is thought that patch-nosed snakes find the latter by odor. These predominantly terrestrial snakes may ascend low shrubs or cactus clumps.

The common name refers to the enlarged, free-edged, rostral (nosetip) scale. It is surmised that this is an adaptation that helps the patch-nosed snakes dig more effectively in loose sand, perhaps to root out buried reptile eggs.

Patch-nosed snakes are oviparous. Based on data from captive examples, the single clutch of 3–10 eggs hatch after about 60 days of incubation. The preferred natural nesting site is unknown.

These desert speedsters prefer fleeing to standing their ground, but if cornered this snake will "S" the anterior part of its body, vibrate its tail in ground debris, and do its best to project an image of a formidable adversary. It will seldom attempt to bite until actually grasped.

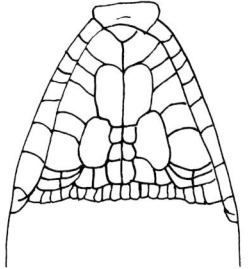 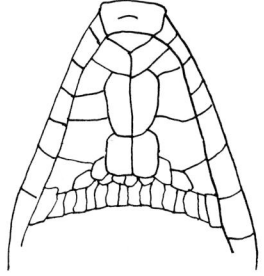

Figure 5.02. Big Bend Patch-nose (left) and Mountain and Texas Patch-noses. Illustration by Patricia Bartlett.

There are 17 rows of smooth scales (occasionally weakly keeled near and above the vent) and a divided anal plate. All species are striped.

Patch-nosed snakes are nonvenomous.

32. Mountain Patch-nosed Snake

Salvadora grahamiae grahamiae

Disposition: Although allied to both the racers and the whipsnakes and certainly well able to bite, patch-nosed snakes are usually somewhat less apt to do so than their larger relatives.

Abundance/Range: This is a common snake, but it is so fast and alert that it is hard to surprise. They are often seen crossing rural desert roadways in the morning hours. This is a snake of moderate elevations, often found at up to about 5,000 feet in foothills and mountains. The mountain patch-nosed snake ranges westward the Big Bend of Texas to central southern Arizona. It is also found in north central Mexico.

Habitat/Range: This is a snake that is entirely at home in habitats as diverse as grassy road verges or cactus-studded desert hillsides. It is not often encountered in urban or suburban settings, but is fairly common in ranchlands and mountain fastnesses. On cloudy days or when preparing to shed its skin, the mountain patch-nosed snake will seek cover beneath debris or in rodent burrows.

Size: The typical adult size is 2–2½ feet, with occasional examples attaining 3 feet in length. Hatchlings are about 9 inches long.

Identifying features: It is the comparative width of the stripes and on what scale rows they occur, as well as the arrangement of the posterior chin scales, that will

32. Mountain Patch-nosed Snake

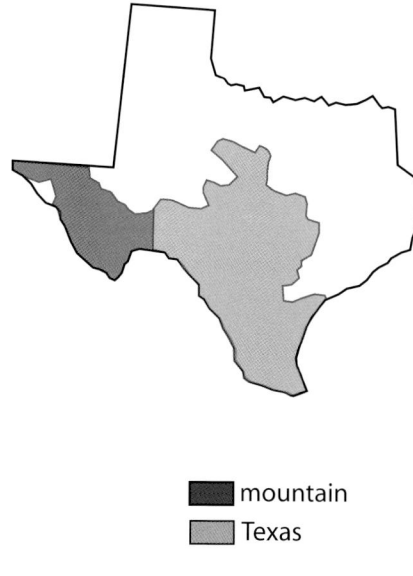

mountain
Texas

identify these snakes to both species and subspecies.

The mountain patch-nosed snake has the fewest stripes of the three types we will discuss. There is a broad, straight-edged, buff-colored vertebral (or back) stripe (normally covering the width of 1 full scale row plus one-half scale row both above and below) that is bordered on each side by an even-edged dark brown to black stripe about as wide as the vertebral stripe. Anteriorly, each dark stripe extends at least to the eye. These are normally the only markings on this snake, but occasionally a very faint light stripe is present (at least anteriorly) on the third scale row up from the ventral plates. The white belly (occasionally suffused with a blush of yellow) is unmarked. Hatchlings are very like the adult in appearance. There is only a single scale between the posterior ends of the large posterior pair of chin scales.

Similar species: Both the Big Bend patch-nosed snake and the Texas patch-nosed snake have well-defined dark lateral stripes. Garter snakes have keeled scales and light rather than dark stripes.

Additional subspecies

33. The Texas Patch-nosed Snake, *Salvadora grahamiae lineata*, is the lower-altitude, more southeastern representative of this species. It is often more colorful, and always more busily patterned than the mountain patch-nose. The Texas patch-nose is found in a broad north-south swath throughout much of central Texas and well into Mexico.

The vertebral stripe of this race is one plus one-half scale row on each side in width. It may be buff, yellow, or orange. The dark brown dorsolateral stripes are 2 scale rows plus one-half row on both top and bottom in width. A thin dark line runs lengthwise on the third row of scales above the belly plates. The belly is buffy to a pale yellow green. The top of the head is usually olive.

As with the mountain patch-nosed snake, there is very little known about the reproductive biology of the Texas patch-nose. A single clutch of 5 eggs, with a

33. Texas Patch-nosed Snake

female in attendance, was found in a rodent burrow beneath a roll of discarded carpet in Texas.

34. Big Bend Patch-nosed Snake

Salvadora hexalepis deserticola

Disposition: Feisty yet harmless describes the attitude and potential of this nonvenomous snake.

Abundance/Range: This rather common snake ranges westward from the Big Bend region of Texas to southeastern Arizona, then far southward along the west coast of mainland Mexico.

Habitat: This is a snake of the arid southwestern flatlands, but it also ranges into the foothills. It is at home on the hot sands of the Chihuahuan and the easternmost reaches of the Sonoran deserts. Look for it amid desert thorn scrub and desert grasses on substrates of sand or gravel.

Size: The most commonly encountered adult size of this desert resident is 24–34 inches. Occasional individuals may attain 38–40 inches. Hatchlings are about 9 inches in length.

34. Big Bend Patch-nosed Snake

Identifying features: Although to say that this patch-nosed snake is the lightest (or grayest) of the three described herein is accurate, the differences are subtle and not apt to be useful unless you have one of the others in hand for comparison.

The Big Bend patch-nosed snake is the only one of the three to have the bottom dark dorsolateral stripe "zigzag" (it follows the edge of each individual scale) rather than straight. The vertebral stripe is orange tan, tan, or pale buff. The sides are gray to pale tan. The 4 dark stripes are brown to almost black with a grayish haze. The belly is often an eggshell hue anteriorly, but may shade to orange or peach posteriorly and subcaudally. The dark dorsolateral stripe extends forward to the eye. The lowest (and thinnest) dark stripe is on the 4th row of scales above the ventral plates.

The pair of large posterior chin scales are entirely separated by 2 or 3 rows of scales.

Similar species: The mountain patch-nosed snake usually lacks the thin lateral line and has a narrow vertebral stripe. The Texas patch-nose has the thin lateral line on the third row of scales above the belly plates, has a narrow light vertebral stripe, and has comparatively wide dark dorsolateral stripes.

Hook-Nosed Snakes and Ground Snakes

Hook-nosed Snakes, genera *Ficimia* and *Gyalopion*

Very little is known about the life history of any of the hook-nosed snakes. All look like diminutive, smooth-scaled hog-nosed snakes. The snakes of the genus *Ficimia* range from southern Texas to northern Central America while those of the genus *Gyalopion* range southward from central New Mexico to central Mexico. In our area of coverage both species are restricted to Texas.

These are small, slow-moving, burrowing, oviparous snakes that eat insect larvae and spiders.

Other than the fact that both have small clutches (1–5 eggs) nothing is known about the breeding strategies of either hook-nosed snake.

Hook-nosed snakes burrow rapidly from sight when frightened. They may writhe spastically from side to side and/or produce a popping sound by extruding and retracting the cloacal tissue through the vent.

Gyalopion has the prefrontal scales in contact, separating the rostral scale from the frontal scale. The inner edges of the prefrontal scales of *Ficimia* are not in contact and the rostral scale touches the frontal scale. The scales are smooth and in 17 rows. Both Texas species have a divided anal plate.

Although the salivary components of the hook-nosed snakes contain toxic components to help them overcome their invertebrate prey, a bite from either of these tiny snakes is of no concern to a human.

35. Tamaulipan Hook-nosed Snake

Ficimia streckeri

Toxicity/Disposition: Toxic to its invertebrate prey, this snake almost never attempts to bite a human.

Abundance/Range: The few researchers to whom we spoke about this snake have indicated that when specifically searched for, it is not hard to find in Texas. Its burrowing habits and nocturnal propensities probably make this snake seem more uncommon than it actually is. It is found from extreme southeastern Atascosa County, Texas, to far south of the international boundary.

Habitat: This is a thorn scrub and grassland snake that often utilizes surface debris such as rocks, piles of moldering vegetation, or old carpets in a dumping site for cover.

Size: Most examples of this tiny almost cylindrical snake are adult at about a foot in length. The record size is 19 inches. Hatchling size is unknown.

Identifying features: This little snake is sandy gray tan or olive tan dorsally and on the upper sides, becoming lighter on the lower sides and venter. The venter is immaculate and white to cream in color. There are many darker brown, narrow dorsal bars (or rows of spots) with uneven edges and a row of smaller, often indistinct spots where the dorsal and ventral color meet. A more prominent small

35. Tamaulipan Hook-nosed Snake

dark spot is present beneath each eye. The upturned rostral scale does not have a medial dorsal keel. The pupils are round.

Similar species: This species and four other eastern and central snakes have noticeably upturned rostral scales. However, the three species of hog-nosed snakes all have keeled body scales. The ranges of the two hook-nosed snakes do not overlap.

36. Chihuahuan Hook-nosed Snake

Gyalopion canum

Toxicity/Disposition: The teeth of this hook-nosed snake increase slightly in length from front to back. The largest rear teeth are grooved. The saliva of this snake contains components that are toxic to invertebrates. It seldom attempts to bite a human and is considered harmless.

Abundance/Range: Once thought rare, this hook-nosed snake is now seen with some degree of frequency after dark as it crosses desert highways. It is found in a broad east-west swath through central Texas to west Texas and southeastern Arizona. A disjunct population exists in the southern panhandle of Texas.

Habitat: This burrowing snake is associated with aridland habitats that support mesquite and other thorn scrub, shin oaks, and juniper. Grasslands are also inhabited. This snake occurs at elevations of about 1,000–6,500 feet. It seeks habitats with surface rocks, vegetation debris, and gravelly soils.

36. Chihuahuan Hook-nosed Snake

Size: Adults typically measure 6–12 inches in length. The record size is $15^{1}/_{16}$ inches. The size of a hatchling is unknown.

Identifying features: The ground color of the Chihuahuan hook-nosed snake often closely matches the color of the soil and rocks in which it burrows. The snake may be pale gray with tan saddles to buff with dark brown saddles. The saddles are much wider than on the look-alike Tamaulipan hook-nosed snake. A row of small lateral spots is often visible. There is little indication of fading to the ventral color on the lower sides. The belly is immaculate white to eggshell. A dark bar connects the eyes over the top of the head. A nape spot is usually well defined and may be connected to the first body blotch. A dark spot is present between each eye. The pupils are round.

Similar species: Hog-nosed snakes have keeled scales. The range of the Tamaulipan hook-nosed snake is not contiguous with the range of this species.

Ground Snakes, genus *Sonora*

Sonora is another genus that is of primarily Latin American distribution. However, one species ranges widely in the western United States and also through Texas and eastward to southwestern Missouri.

This is a tiny, secretive, burrowing snake with more color phases than are easily described. It is a completely harmless insectivore that can be very abundant, yet is so secretive its existence in a given area may not even be suspected.

Only a single clutch of 2–6 eggs is laid each summer. The incubation is a few days on either side of 2 months.

The same death shamming that is so well known in hog-nosed snakes has also been documented in ground snakes. The snake first writhes about, tongue lolling, then rolls onto its back, mouth open, tongue still extended. The only sign of life in a snake "playing dead" is a rolling back into a belly-up position if righted.

Prey items include scorpions, centipedes, beetle larvae, crickets, and spiders.

A loreal scale is present anterior to the eye. This separates the second supralabial scale from the prefrontal scale. This is important when trying to differentiate some phases of this snake from the various crowned snakes (all of which lack a loreal scale). This snake may have either 14 or 15 rows of smooth scales at midbody; the anal plate is divided.

This is a nonvenomous snake.

37. Great Plains Ground Snake

Sonora semiannulata semiannulata

37a. Great Plains Ground Snake

37b. Great Plains Ground Snake

37c. Great Plains Ground Snake

37d. Great Plains Ground Snake

37e. Great Plains Ground Snake

Disposition: This snake can seldom be induced to bite.

Abundance/Range: The Great Plains ground snake is a widespread, abundant species, but may be of local distribution. It ranges westward from southwestern Missouri and eastern Texas to northwestern Nevada and southeastern California. Disjunct populations occur elsewhere in the west, and the contiguous range continues southward into Mexico.

Habitat: The ground snake is associated with sandy grasslands and pastures, sandy rock-strewn road verges, or sandy deserts with surface rocks (in the southwest), dry washes, and mesquite thickets, and riverbanks. It hides beneath rocks and debris, remaining near the ground surface in the springtime, but apparently burrowing down into yielding soils at other times of year or during drought. Ground snakes are occasionally very common near water sources such as small streams or stock watering tanks.

Size: This is a small and easily overlooked snake. Most adults are only 8–12 inches long, with a record of 18⅞ inches. Hatchlings are about 3¾ inches long.

Identifying features: In the case of this snake, a picture, or a series of pictures actually, is worth a thousand words. This is one of America's most variably colored and patterned snakes. Photos 37a–37e demonstrate some of the color variations, but by no means exhaust all possible combinations. One of the few constants on this small snake is the belly color, from plain white to cream. The subcaudal color may be brighter. The head is slightly wider than the neck. The presence of a loreal scale will differentiate the ground snake from some of the members of the genus *Tantilla*. However, the loreal scale may be partially fused to surrounding scales, confusing the issue. The smooth scales are in 13–15 rows; the anal plate is divided.

Similar species: The various black-headed and flat-headed snakes lack a loreal scale. Brown snakes have keeled scales.

Comments: Because of the immense variation within populations of this snake, there has been a tendency for researchers to not recognize subspecies of the ground snake. However, it has been suggested that the south Texas examples of the ground snake, a population with a comparatively standardized suite of

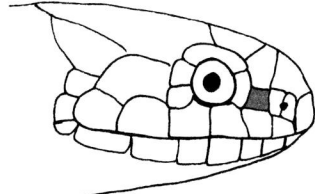

Figure 5.03. Ground Snake, showing preocular. Illustration by Patricia Bartlett.

37f. Taylor's Ground Snake

color and pattern characteristics, should be differentiated. For these the scientific name of *Sonora semiannulata taylori* is proposed. The common name that has been used for this form is Taylor's Ground Snake.

Taylor's ground snake is gray to medium brown above with a short, thin, darker horizontal mark on each dorsal and lateral scale (see photo 37f). There is a tendency for the head to be a little darker than the body. The unmarked belly is a very pale yellow. There are 13 or 14 rows of dorsal scales. Differentiate this snake from the look-alike flat-headed snake by the presence of a loreal scale.

Black-headed Snakes and Crowned Snakes, genus *Tantilla*

In the Southeast the snakes of this genus are diminutive burrowers. Some species in the south central states are larger, but are equally secretive. Most are characterized by a black crown and, sometimes, a black nape. The head and neck color may or may not be separated by a light band.

The unicolored dorsum is often the color of the substrate on which these snakes are found. There is no loreal scale, an important fact when trying to differentiate black-headed snakes from ground snakes. The eyes are proportionately small and the head and neck are the same diameter as the body.
These are snakes of sandy, well-drained soils. They may occasionally be found beneath flat stones and building debris, but are usually in burrows. Some species are common but others are uncommon to rare.

The scales of these oviparous snakes are in 15 rows; the anal plate is divided. Traditionally, the shape of the posterior edge of the dark crown and presence or absence of a light collar have been considered the most reliable identifying characteristics. While these markings must be considered, there is actually

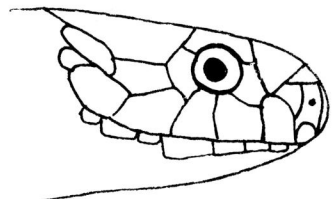

Figure 5.04. Tantilla. No loreal scale. Illustration by Patricia Bartlett.

much variation in crown characteristics even within a subspecies. Rely also on range to identify these snakes.

The reproductive biology of the snakes in genus *Tantilla* is unusual, even if of no consequence to a field observer. These snakes have two functional ovaries, but only a single functional oviduct. Eggs formed in the left ovary are shunted into the functional right oviduct for deposition. Because spermatogenesis occurs at a different time of year than ovulation, sperm storage is necessary. Following mating, sperm is retained in the nonfunctional, vestigial section of the left oviduct until fertilization of the ripe ovum occurs.

A single clutch of 1–5 eggs is laid annually by both small and large species and incubation takes about 60 days.

Black-headed snakes are nervous, move quickly when uncovered, and may thrash from side to side in evasive attempts. Although harmless to humans and reluctant to bite, crowned snakes possess toxic saliva. Centipedes seem to be the primary prey items of many species of these snakes. The toxins contained in the saliva of the snakes quickly overcome and immobilize the multisegmented, biting chilopods. Insects and their larvae are also eaten.

38. Mexican Black-headed Snake

Tantilla atriceps

Toxicity/Disposition: This snake has weak venom that helps it overcome its arthropod prey. It does not bite.

Abundance/Range: Although not uncommon in north central Mexico, to date this species has been found in the United States only in Duval and Kleberg counties in south Texas.

Habitat: In the United States the Mexican black-headed snake is known to occur only in desert brushland. It is probable that it secretes itself beneath moisture-retaining flat rocks or debris, at least in the spring and early summer. It is a persistent burrower.

Size: This snake attains a maximum size of about 9 inches.

Identifying features: The dorsal coloration of this tiny snake is tan to pale brown. The belly is almost white along its outer edges and pink midventrally. The dark cap seems to balance delicately on the top of the head, wrapping downward posterior to the eyes only to the upper edges of the supralabial scales and extending back from the parietal scales only one or two scale rows. The posterior edge of the dark cap is almost straight across, not pointed or rounded. There is no light collar. There are 7 supralabial (upper lip) scales. The first lower labial scale on each side usually meets at the midline of the chin.

Similar species: The cap of the Plains black-headed snake extends rearward in a rounded point. The flat-headed snake does not have a distinct dark cap. On the southwestern black-headed snake, the first chin scale on each side usually does not touch on the midline of the lower jaw.

38. Mexican Black-headed Snake

39. Southeastern Crowned Snake

Tantilla coronata

Toxicity/Disposition: Very mildly venomous. Although the saliva of this snake is mildly toxic, the snake is entirely harmless to humans, and usually cannot be induced to bite.

Abundance/Range: Because of its persistent burrowing habits, actual population statistics are difficult to assess. However, this is probably not an uncommon snake. It is found throughout most of the Southeast, from northern Kentucky and central Virginia to eastern Louisiana and the Florida panhandle.

39. Southeastern Crowned Snake

Habitat: Habitats are many and varied. Soil moisture can vary from moderate (not wet, though) to very dry and the ground cover may be woodlands, scrubland, or grassland—the latter including urban fields and yards.

Size: This tiny snake is often hardly larger than a big earthworm, measuring 7–12½ inches as an adult. Hatchlings are about 3 inches in length.

Identifying features: Range alone will identify this black-headed snake. The dorsal color is tan to reddish tan, usually closely mimicking the color of the soil where it dwells. The belly is an opalescent white to pale yellow or pink. The rear of the black crown is usually gently curved, but may have short rearward extensions following the sutures of the parietal shields. The black of the head may reach the mouth line, or a light supralabial streak may be present. A light collar followed by a large black nape-blotch is present.

Similar species: Use range to identify this species. This is the only crowned snake throughout most of the Southeast. Range alone will identify it.

40. Trans-Pecos Black-headed Snake

Tantilla cucullata

Toxicity/Disposition: Despite its rather large size, the weak venom of this black-headed snake appears to be of no medical significance to humans. It will bite if restrained.

Abundance/Range: Once thought to be quite rare, in recent years night hunts on desert roads have produced numbers of these snakes. The exact range remains unclear, but this snake seems to be of rather wide distribution (or at least of wider distribution than once thought) in the Big Bend region of Texas.

Habitat: This is a snake of rocky hillsides and slopes, rocky river banks, rocky canyons, and even rocky pastures and grasslands. However, with this now said,

it should be noted that the trans-Pecos black-headed snake is almost never found except when it is surface-active on cool, preferably overcast, nights.

Size: This is the largest species of black-headed snake in the United States. Adults are usually 10–15 inches in length. The record size is 25⅝ inches. Hatchling size is unknown.

Identifying features: The ground

40a. Trans-Pecos Black-headed Snake, black-headed morph

40b. Trans-Pecos Black-headed Snake, collared morph

color of this snake varies from a sandy olive tan to a reddish brown. The belly is white. The shape and extent of the black cap is variable. At one extreme the head and chin, and the first 5 or 6 rows of scales, including the first 5 or 6 ventral scutes, are entirely black. At the other extreme a solid white collar, about 2½ scale rows in width, is present at the back of the head. Between these extremes the white collar may be incomplete vertebrally, the two halves being separated by ½–5 rows of white scales. In some areas of the range, such as in the mountain ranges of Big Bend National Park, the cap configuration seems influenced by altitude. Collared examples usually have a light spot on the rostral scale, another on the first supralabial, and a third (the largest) on supralabial number 5.

Similar species: Size alone will identify the adults of this species from other black-headed snakes. The rear edge of the black cap of the Plains black-headed snake is rounded or pointed. The southwestern black-headed snake has a light upper lip and chin.

41. Flat-headed Snake

Tantilla gracilis

Toxicity/Disposition: This tiny snake has weak venom that helps it overcome its arthropod prey. It is reluctant to bite and a bite is of no medical consideration to humans.

Abundance/Range: This is a common to abundant snake. It ranges southward from northeastern Kansas and central Missouri to southern Texas and Coahuila, Mexico.

Habitat: The flat-headed snake may be found beneath many kinds of surface debris. Vegetation, rocks, or human-generated debris is all utilized. This species is a habitat generalist, occupying grasslands, treed bottomlands, mixed upland forests, and rocky juniper brakes. It may be found from isolated wooded canyons to desert brushlands, vacant fields, and backyards. Like other members of the genus, this snake is a persistent burrower and not often surface active.

Size: Adults are 6–8 inches in length. The record size is only 9⅞ inches. Hatchlings are about 2¾ inches in length.

Identifying features: The dorsal coloration of this tiny snake is tan to pale brown or brownish red. The belly is almost white along its outer edges and pink

41. Flat-headed Snake

midventrally. The cap coloration is often only barely darker than the dorsal color and may not have a well-defined posterior edge. However, if able to be determined, the posterior edge of the cap is either straight across or may even be slightly concave. There is no light collar. The cap wraps downward posterior to the eyes only to the upper edges or centers of the supralabial scales. There are 6 supralabial (upper lip) scales.

Similar species: Other members of this genus have black caps or heads. The ground snake has a loreal scale. Brown snakes and earth snakes have at least some keeled scales.

Comments: In the northern portion of its range, the flat-headed snake may burrow deeply to undergo its winter hibernation (also called brumation). However, further south, where hard freezes are less likely, this snake has been found coiled quietly beneath surface debris in all winter months.

42. Southwestern Black-headed Snake

Tantilla hobartsmithi

Toxicity/Disposition: This tiny, secretive snake has weak venom but a bite would be of no medical consideration to humans.

Abundance/Range: This black-headed snake is commonly seen throughout its extensive range. It occurs over virtually all of southwestern Texas as well as in adjacent New Mexico and Mexico, and in disjunct populations in Arizona, California, Colorado, and Nevada.

Habitat: The southwestern black-headed snake occupies a wide variety of habitats. It may be found beneath vegetation debris, rocks, or human-generated debris in isolated wooded canyons, desert brushlands, and roadside parks. Like other members of the genus, this snake is a persistent burrower.

42. Southwestern Black-headed Snake

Size: This snake attains a maximum size of about 9 inches. Hatchlings are about 3¼ inches in length.

Identifying features: The dorsal coloration of this tiny snake is tan to pale brown. The belly is almost white along its outer edges and pink midventrally. The black of the cap stops just posterior to the large parietal scales. The posterior edge of the dark cap is almost straight across, not pointed or rounded. There is no light collar. The cap wraps downward posterior to the eyes only to the upper edges or centers of the supralabial scales. There are 7 supralabial (upper lip) scales. The first lower labial scale on each side usually does not meet at the midline of the chin.

Similar species: The cap of the Plains black-headed snake extends rearward in a rounded point. The flat-headed snake does not have a distinct dark cap. On the Mexican black-headed snake, the first chin scale on each side usually touches on the midline of the lower jaw, but range alone will differentiate the southwestern black-headed snake from the Mexican black-headed snake.

Comments: Unlike most other species in this genus, which prey largely on centipedes, this species eats many butterfly/moth larvae and spiders.

43. Plains Black-headed Snake

Tantilla nigriceps

Toxicity/Disposition: This species, like most other black-headed and crowned snakes, can seldom be induced to bite.

43. Plains Black-headed Snake

Abundance/Range: This is a common black-headed snake. It ranges southward from southwestern Nebraska and northern Colorado to southern Texas and central Arizona. It is also found in Mexico.

Habitat: The Plains black-headed snake is often found beneath many kinds of surface debris, including vegetation debris, rocks, and trash. Although primarily a grassland species, the Plains black-headed snake also may be found in isolated wooded canyons, desert brushlands, and roadside parks. Like other members of the genus, this snake is a persistent burrower and not often seen above ground.

Size: Adults are 8–12 inches in length. The largest known specimen was only slightly more than 15 inches long. Hatchlings are about 2½ inches in length.

Identifying features: The dorsal coloration of this tiny snake is tan to pale brown. The belly is almost white along its outer edges and pink midventrally. The black of the cap extends in a point or a curve well beyond the large parietal scales. There is no light collar. The cap wraps downward posterior to the eyes only to the upper edges or centers of the supralabial (upper lip) scales. There are 7 supralabial scales. The first lower labial scale on each side usually meets at the midline of the chin.

Similar species: The cap of the southwestern and the Mexican black-headed snakes follows the curvature of the head scales and does not extend into a point or curve on the nape. The flat-headed snake does not have a distinct dark cap.

Comments: This is another of the black-headed snakes that eats a preponderance of Lepidoptera larvae, beetle larvae, and spiders. Centipedes are occasionally eaten.

44. Rimrock Crowned Snake

Tantilla oolitica

Toxicity/Disposition: Like others in this genus, the saliva of the rimrock crowned snake is mildly toxic. However, the snake is entirely harmless to humans and usually cannot be induced to bite.

Abundance/Range: The population statistics of this snake are unknown, but it is thought to be a rare species. It is found only in Dade and Monroe counties, Florida, and is protected by state law.

Habitat: Despite extensive development in Miami-Dade and Monroe counties, this little snake continues to turn up on occasion. Examples have recently come from some of the more southerly of the Upper Keys. Rimrock crowned snakes have been found in piles of damp discarded clothing as well as beneath rocks.

44. Rimrock Crowned Snake

This is the only crowned snake known from southeastern Florida and the Keys. **Size:** This tiny snake is often hardly larger than a big earthworm, with an adult size of only 7–9 inches.

Identifying features: On mainland examples, the black of the head extends well beyond the head scales onto the nape. There is no light collar. Examples from the Keys have a dark head, a poorly defined lighter collar, and a nape ring that is darker than the head. The body color varies from pale whitish tan to a rather rich tannish brown. The belly is white.

Similar species: Use range to identify the rimrock crowned snake. It is the only species of this genus to occur in Dade and Monroe counties, Florida.

45. Peninsula Crowned Snake

Tantilla relicta relicta

Toxicity/Disposition: This snake usually cannot be induced to bite.

Abundance/Range: Its persistent burrowing habits make actual population statistics of the peninsula crowned snake difficult to assess. However, it seems to be a common snake.

There are four disjunct ranges, all restricted to Florida. The largest range extends southward from southern Clay County to Highlands County to Hillsborough County. A population also occurs on Cedar Key (and other nearby Keys) in Levy County, Florida. A third population occurs near the coast in Sarasota

45. Peninsula Crowned Snake

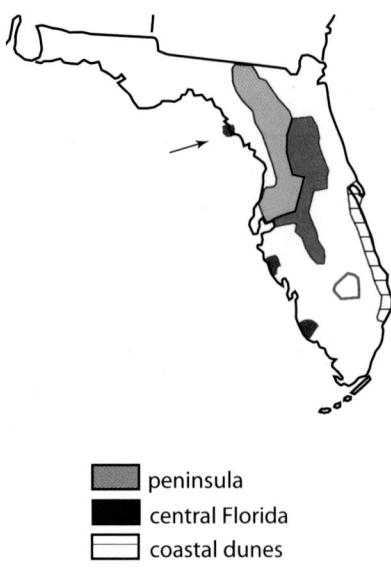

peninsula
central Florida
coastal dunes

County, and a fourth in Charlotte, Lee, and Collier counties.

Habitat: This crowned snake is often raked from just beneath the surface of the soil when the weather is warm and damp, but burrows deeply to avoid excessive heat, cold, or drought. It may also be found beneath surface debris.

Size: A large adult of this tiny species is only 7–9 inches long. Hatchlings are about 2¾ inches in length.

Identifying features: The black of the head is usually separated from the black of the nape by a light collar, but this can be variable. The body color is sandy tan to reddish brown, usually closely simulating the color of the soils on which the snake is living. The belly is white.

Similar species: Use range as an identification aid. Where this form abuts the central Florida crowned snake, specimens of intermediate appearance are often seen. The peninsula crowned snake usually has a light collar that separates the black of the head from that of the nape, while the central Florida crowned snake lacks the light collar.

Comments: In normally damp, warm weather we have commonly found this snake a few inches beneath the sand surface at various locales on the Lake Wales Ridge. At times of drought and in cold weather we have found none and surmise that the snake burrows to follow the moisture line and to avoid temperature extremes.

Additional subspecies

46. The Central Florida Crowned Snake, *Tantilla relicta neilli*, has the black of the head extending well beyond the head scales onto the nape. There is usually no light collar; in its place might be an upward notch in the black pigment at the lower sides of the rear of the head. The body color is sandy tan to reddish brown, usually closely matching the color of the soils on which the snake is living. There are 15 rows of smooth scales and a divided anal plate. The range of this fossorial snake arcs inland and upward from Hillsborough County to Polk

46. Central Florida Crowned Snake

County to Madison County. Found virtually to the Georgia state line, the range of *T. r. neilli* does not seem to extend into that state.

47. The Coastal Dunes Crowned Snake, *Tantilla relicta pamlica*, is the eastern-most race of this species. It is found along Florida's east coast from Palm Beach County north to northern Brevard (and perhaps southern Volusia) County. The rear of the crown is gently curved, but may have short rearward exten-sions following the sutures of the parietal shields. The medial projection may be especially prominent and may narrowly break the light collar, connecting the black of the head to the black nape patch. The body color is sandy tan to reddish brown.

47. Coastal Dunes Crowned Snake

Lyre Snakes, genus *Trimorphodon*

The lyre snakes have derived their common name from the lyrelike pattern on the top of their head. The single species found north of Mexico has several subspecies. Only one, the Texas lyre snake, occurs within the scope of this book and, in the East, is restricted to West Texas. The Texas lyre snake lacks a distinctive head pattern.

The lyre snakes have a head that is much wider than the neck, vertically elliptical pupils, 23 rows of smooth scales, and a divided anal plate.

Little is known about the reproductive biology of the Texas lyre snake. A 28–inch captive female laid 6 eggs from which, after an incubation of 77 days, hatchlings approximately 8½ inches in length emerged.

Because of the remoteness of its range, the hostility of its preferred habitat, and its nocturnal habits, the lyre snake is not a commonly encountered species. The few that are found are most often seen on spring or summer nights as they cross one of the few roadways that traverse their habitat. Lyre snakes usually move rather slowly but are capable of bursts of speed if they are frightened. If cornered they assume a loose S shape and will strike (often, it seems, with mouth closed) and vibrate their tail. This snake species feeds predominantly on lizards, but may accept small snakes and nestling rodents and birds as well.

Lyre snakes are rear-fanged, mildly venomous snakes. Their venom quickly immobilizes the lyre snake's lizard and snake prey. It seems less effective on small mammals and nestling birds. Bitten humans have experienced localized itching, numbness, and redness. These snakes have catlike elliptical pupils, and seem to have excellent night vision. They are almost exclusively nocturnal in their activity patterns.

48. Texas Lyre Snake

Trimorphodon biscutatus vilkinsonii

Toxicity/Disposition: The Texas lyre snake, a mildly venomous snake species, often strikes when on the defensive, but is somewhat less apt to actually bite. The makeup of the venom is rather similar to that of the dangerously venomous elapine snakes, but it is not considered life threatening to humans, or even of great medical significance by most researchers.

Abundance/Range: The Texas lyre snake has a restricted distribution in Texas and is not always easy to find. It is protected by the state of Texas and should not be collected without a specific permit. It occurs along the Rio Grande from Brewster County in the Big Bend area to central New Mexico and north central Mexico.

48. Texas Lyre Snake

Habitat: This is a snake of gravelly hillsides, boulder fields, and heavily wooded montane canyons as well as aridland thorn scrub. It is most often found while one is driving the back country roadways at night. This snake has been most intensively studied in Big Bend National Park, the Franklin Mountains, and similar West Texas sites.

Size: The lyre snake is adult at a length of 2–2½ feet. The record size is 41 inches. Hatchlings are 8½ inches long.

Identifying features: This pretty, slender, bigheaded snake is attractive but not brightly colored. It occurs in light grayish tan, a dark olive gray, and an interim color phase. All have a similar pattern, but a differing ground color. The dark blotches, which begin on the nape, number about 25 on the body and 10 on the tail. The body and anterior tail blotches are broadest dorsally, where they have light centers and light outer edges. The more distal tail blotches are narrow and lack light centers, but still retain light edging. There are much smaller dark blotches positioned on the lower sides about midway between the lower extensions of the dorsal saddles. Except for where the lower dark blotches encroach on the ventral scales, the belly is off-white to pale tan.

The dorsal color of Texas lyre snakes of the dark variation is olive brown. The light-phase snakes are pale gray, and those in between are a pale grayish buff to rich buff. The top of the head, so prominently marked in other subspecies of lyre snakes, may be without contrasting markings, have a trio of small spots, or bear a barely discernible spearpoint. A dark bar may be visible between the eyes. Hatchlings usually have a ground color of pale gray, dorsal spots with less prominent light centers, and edging to the dorsal blotches.

Similar species: The Texas lyre snake (especially as a juvenile) very closely resembles some color morphs of the gray-banded kingsnake. Both are found in the Big Bend region of Texas and both are nocturnally active. However, the gray-banded kingsnake has round pupils in its large eyes and often has some red or orange in the dorsal color scheme. Copperheads are a distinctive tan with terra-cotta bands that are broader on the sides than on the back. Night snakes have dark dorsal spots without light centers and an elongate dark blotch on each side of the face and neck.

Black-striped Snakes, Night Snakes, Cat-eyed Snakes, and Pinewoods Snakes, subfamily Dipsadinae

The Dipsadinae are of primarily Neotropical distribution. In fact, only four species (each in a separate genus) occur in the United States. Of these, three species occur in Texas (and the west) and one is restricted to the Southeast.

The snakes of this family are grouped together because of hemipenial morphology. Some members of the family autotomize (break) their tail rather readily. Unlike the tails of many lizards, the tails of these snakes do not regenerate. All members of this subfamily have enlarged teeth at the rear of the upper jaw, and weak venom to help them overpower their prey of lizards, frogs, and salamanders (and possibly invertebrates).

The black-striped snake at home

There are many smaller snakes that are so secretive that they are usually seen only when a piece of ground-surface debris is overturned. Old boards, roofing tin, and carpets provide ideal cover for these species; even larger, normally active snakes may also seek the protection of ground cover when they are preparing to shed their skin. Whether in the desert or in Florida, snakes seek out this sort of cover.

We were in southern Texas looking for herps to photograph. We knew our best bet for success would be debris that we could lift and dig through, so we were naturally excited when we happened upon a roadside dumpsite containing boards, tin, and carpet pieces.

As is often the case, we found little at the largest trash pile, and had moved to the outlying pieces of plyboard and carpet. Lifting the boards produced nothing. The ground was cracked and dry beneath them. The first few pieces of carpet were also devoid of skulking beasts beneath, but the carpet seemed ideally to be just on the damp side of being completely dry.

Finally we had worked our way to the last piece of cover. It was a remnant of carpet about a yard square that lay mostly in the shade of a small palmetto.

It was beneath this square of carpet that we found the prizes of the trip. As the damp, discarded remnant was rolled back, two little black-striped tan snakes were left exposed. For a few moments both lay quietly entwined; then, as if simultaneously realizing they had suddenly become vulnerable, both snakes began probing the ground for escape burrows.

This was our introduction to the black-striped snake, a primarily Latin American species that enters our area only in the lower Rio Grande valley of Texas. This species, and the northern cat-eyed snake, also of South Texas and Latin America, are the two most beautifully colored species of this subfamily in the United States (representatives of the family do not occur in Canada).

Adding cover works to lure these snakes to one's own back yard. We have seeded our Florida back yard with pieces of board just in the hope that small native snakes will take up residency. Once the boards have weathered in, they have been rather productive.

It is not uncommon to overturn one of the boards and find one or more pine woods snakes. Some individuals are recognizable because of blemishes or healed injuries. One such, a large female, has her tail broken off just behind the vent. For three summers now she has been found beneath a particular large board that lies in patchy sunlight. Hers is the only board in the yard where we do not regularly find numbers of greenhouse frogs underneath. Could this be merely coincidence, or do the tiny frogs--interlopers at virtually all other areas of cover-- recognize that particular board as a palace of doom?

Black-striped Snakes, genus *Coniophanes*

This Latin American species barely enters the United States in the vicinity of Brownsville, Texas. There it is most common, but secretive, beneath debris in yards and parks. It is most active at dusk (a crepuscular activity pattern), but also apt to be seen on irrigated lawns well after nightfall.

Like other members of this genus, the Tamaulipan black-striped snake elevates and writhes its tail when disturbed. Speculations differ regarding the reason for this behavior and for the loss of tail in this snake. Some researchers feel that the tail is writhed as a diversionary tactic and if seized by a predator is autotomized (i.e., breaks free). Others have attributed the preponderance of foreshortened tails seen on black-striped snakes to "tail rot."

We feel that the first theory is the correct one. On two separate occasions the tails of snakes that I tried to stop from escaping beneath logs came off in my hand. On these two occasions, at least, the loss had nothing whatever to do with tissue necrosis.

The snakes in this genus produce venom that is conducted via grooves in the enlarged rear teeth into bite wounds. This is considered a primitive venom delivery system. The venom is relatively effective against the amphibians and small reptiles that are predominant in the snake's diet but the effects on warm-blooded (endothermic) animals are less well known. One envenomation of a human by a black-striped snake produced lingering and painful effects. Despite this, the black-striped snake is not considered dangerous to humans.

The members of this genus are all egg layers. It is probable that in Texas this species lays only a single clutch of eggs annually. Of the two clutches reported, one contained 4 eggs and the other 5 eggs. The incubation duration is about 60 days.

The pupil is round. The 19 rows of scales are smooth and the anal plate is divided.

49. Tamaulipan Black-striped Snake

Coniophanes imperialis imperialis

Toxicity/Disposition: Mildly venomous. This snake can, but almost never does, bite.

Size: Most adults of this shiny little snake are 12–18 inches in length. The record size is a mere 20 inches. Hatchlings are about 6 inches long.

Abundance/Range: Although still considered rare, and protected by law in Texas, in recent years this snake seems to have increased in numbers. It is widespread in Latin America but restricted in distribution in the United States to extreme southern Texas.

Habitat: The black-striped snake is most often found in yards and fields in the area of Brownsville, Texas, beneath vegetation debris or other moisture-retaining surface litter. It is often found in dumps beneath discarded carpet, moldering palm trunks, and piles of palm fronds.

Identifying features: This is one of the most distinctively marked and easily identified snakes of the lower Rio Grande valley of Texas. Dorsally this snake has 3 well-defined dark stripes and 2 light stripes. The dark vertebral stripe is about half as wide as the 2 dark lateral stripes and the 2 light dorsolateral stripes.

49. Tamaulipan Black-striped Snake

The top of the head is dark. A very thin light line projects a few scales posterior (and occasionally anterior) to each eye. The belly and subcaudal scales are orange red.

Similar species: None. Patch-nosed snakes have 2 to 4 dark stripes and a light vertebral stripe. Garter and ribbon snakes have greenish bellies and light vertebral stripes.

Night Snakes, genus *Hypsiglena*

There is only a single subspecies of night snake in the south central United States, but a number of races range through the American West and Mexico. All are relatively small, all are essentially nocturnal, and all seem to be common, especially in areas where rocks and boulders are a prominent feature of the landscape. They are often encountered crossing roadways in desert or semi-aridland habitats. These nocturnal snakes have a slender neck and a wide head. They are quietly colored in the hues of the sands, rocks, and aridland vegetation among which they dwell. Even when unceremoniously grasped the snake can seldom be induced to bite. Some frightened specimens may roll into a ball with their head tucked into the coils.

Lizards of many species are the predominant prey items, but tiny snakes, reptile eggs, amphibians, and insects are also eaten. Although this snake is considered mildly venomous, it has recently been noted that, despite a primitive delivery system, the venom of the night snake rapidly overcomes ectothermic prey.

The night snake lays small clutches of relatively large eggs. Although up to 9 eggs may be laid, 2–5 is the normal clutch size. It is thought that in the United States only a single clutch is laid annually (gravid female night snakes have been found very early in the year so there is a possibility that two clutches are laid by some females). Incubation lasts about 2 months.

Night snakes have vertically elliptical pupils and 21 rows of smooth scales; the anal plate is divided.

50. Texas Night Snake

Hypsiglena torquata jani

Toxicity/Disposition: Most authorities class this snake as a harmless species because it is small, has a primitive venom-delivery system of ungrooved or weakly grooved enlarged teeth at the rear of the maxillary bone, and is reluctant to bite. However, the venom has proven quite potent to prey lizards and care should be used when handling large examples of this snake.

Abundance/Range: This is a common snake but because it is almost entirely nocturnal, its presence is often overlooked. This is the only night snake to be found in the eastern or central states. It ranges widely through most of the western two-thirds of Texas, much of New Mexico, southeastern Colorado, western Oklahoma, and extreme south central Kansas.

50. Texas Night Snake

Habitat: The Texas night snake may be seen crossing roadways from dusk until well after dark. Typically it prefers rocky and gravelly aridland and semi-aridland habitats but may be found in sandy deserts and grasslands as well. Several individuals have been found in east Texas and along the Gulf Coast of central southern Texas where habitats lack the rocky aspect considered typical. Along the southeastern coast of Texas the night snake is associated with pine or pine-hardwood woodlands in the north and sandy pinelands and thorn scrub along the coast. This little snake occurs in some mountain ranges at elevations of up to 5,200 feet.

Size: Most examples of the Texas night snake are adult at 12–15 inches. The record size is 20 inches. Hatchlings are about 6 inches long.

Identifying features: Although there are two rather standardized patterns for the Texas night snake, the actual intensity of color and contrast of pattern against the ground color is highly variable. The ground color may vary from a light sandy olive tan to gray and olive gray. The saddles may be light olive brown to a dark brownish black. The largest and often best defined dark markings are a pair of dark blotches that begin as an oblique bar posterior to the eye, then widen and extend on to the sides of the neck. A dark, elongate, medial blotch

begins on the back of the head and extends far onto the nape, often connecting to the first medial dorsal blotch. The medial pattern may be of simple rather rounded saddles, or the saddles may be broken and offset along the backbone. Two rows of smaller light blotches alternate along each side. The belly is white or very pale yellow. The lightest colored examples are found in the most westerly regions. The pupils are vertically elliptical.

Similar species: There are several snakes that have a pattern and color superficially like that of the Texas night snake, but all but the night snake have round pupils. Among these look-alike species are glossy snakes (see accounts 53, 54, and 55), the western corn snake relatives (accounts 62 and 63), juvenile Baird's rat snakes (account 60) and both the Tamaulipan and the Chihuahuan hook-nosed snakes (accounts 35 and 36).

Cat-eyed Snakes, genus *Leptodeira*

Like many other rear-fanged snakes, the cat-eyed snake has elliptical pupils. It is from the pupils—narrow slits in bright light but wide and almost round under low-light conditions—that the common name has come.

The (apparently) weak venom is conducted into bite-wounds via grooves in the enlarged rear fangs.

This snake is found only in extreme southern Texas, where its semi-aridland thorn brush habitats are rapidly being replaced by oilfields and agriculture.

The northern cat-eyed snake is protected by law in Texas.

A single clutch of 4–12 eggs is laid each summer. The incubation time is somewhat longer than it is for many other eastern snakes, often nearing a full three months.

It searches the vegetation at edges of ponds and resacas after nightfall for amphibians or their eggs. It also eats lizards, and has been reported to eat an occasional fish. Nestling rodents may rarely be accepted. It is so slender that when among tendrils this snake is easily overlooked.

If unable to escape a perceived threat, the cat-eyed snake may "S" the anterior of its body and, with its flattened (thus wider and more threatening appearing) head held well above the ground, may strike several times in fairly rapid succession. It is usually reluctant to actually bite. Alternatively, it may simply coil tightly, hide its head beneath its coils, and allow itself to be handled indiscriminately.

At midbody the scales of this oviparous snake are in either 21 or 23 rows. The anal plate is divided.

51. Northern Cat-eyed Snake

Leptodeira septentrionalis septentrionalis

Toxicity/Disposition: This snake is one of several rear-fanged species that range northward from Latin America into southern Texas. The venom glands are relatively large, the rear fangs are grooved, but the venom seems to be of comparatively low toxicity. The cat-eyed snake often does not rely on its venom to overpower its prey, instead swallowing the frogs and tadpoles it prefers alive. Larger prey is envenomated and held until struggling ceases. The cat-eyed snake is not usually inclined to bite defensively.

Abundance/Range: In the United States this cat-eyed snake occurs only in Cameron, Kennedy, and Starr counties of Texas. It is widespread and abundant south of the international boundary.

Habitat: In Texas this is a snake of arid-land scrub—a rapidly disappearing habitat type—and in these it is most common in the vicinity of ponds and stock tanks in which amphibians breed.

Size: The northern cat-eyed snake is adult at 1½–2½ feet in length. A very occasional individual may reach a full 3 feet. Hatchlings are about 9 inches long.

Identifying features: This is a beautiful, prominently marked, and easily identi-

51. Northern Cat-eyed Snake

fied snake. The ground color is tan, buff, pale gold, or pale orange yellow. The pattern of large, dark brown dorsal saddles is quite distinctive. There is a rearward-directed dark spearhead on the crown and a thin dark line posterior to each eye. The eyes are large and have vertically elliptical pupils. The head is much wider than the neck. The belly is pale orange anteriorly but becomes brighter posteriorly. Each belly scale has dark pigment on its trailing edge. Scales on the underside of the tail (subcaudal scales) may be pale but are often bright orange. Hatchlings are similar to the adults in pattern, but have stronger, more contrasting colors.

Similar species: The Texas night snake has a grayish ground color and proportionately smaller dorsal blotches. The brown-banded morph of the ground snake lacks the pattern contrast of the cat-eyed snake, has a narrow head and round pupils.

Pine Woods Snake, genus *Rhadinea*

Although *Rhadinea* is well represented in Latin America, only a single species of this genus occurs in the United States. This is the pine woods snake, a species of the southeastern and Gulf Coastal plains. The pine woods snake is small, secretive, and oviparous. This snake is typically found in pine flatwoods, where it hides beneath the bark or decomposing fallen pine trunks. It also occurs in damp backyards and old fields, and can be found beneath urban litter, mats of vegetation, and roadside debris. In old texts, this species was referred to as the yellow-lipped snake.

We have never found this snake out and crawling. All the ones we've seen, day or night, have been beneath debris. In most cases the snakes were not particularly nervous when uncovered and allowed themselves to be lifted and handled without showing defensive belligerence. However, one example flattened its head and struck repeatedly but did not bite.

The term "secretive" fits many snakes rather well, but seems especially suitable for the pine woods snake. In fact, the Neotropical species of this genus are referred to as leaf-litter snakes, for their evolved niche is under and amid the leaf litter of the rainforests.

A clutch of 3 eggs, found beneath a board in June, hatched after 51 days. How long they had been laid before being found is unknown. Up to 4 eggs in a clutch have been documented.

Small amphibians and lizards are eaten after being immobilized by this snake's mildly toxic saliva. It is considered harmless to man.

52. Pine Woods Snake

Rhadinea flavilata

Toxicity/Disposition: The modified saliva of this snake is harmless to humans but effectively quiets its prey. This snake can seldom be induced to bite.

Abundance/Range: Because it is so secretive it is difficult to assess the actual abundance of the pine woods snake, but it is probably not rare. This species occurs in several disjunct populations. It is found in coastal North Carolina and South Carolina, near Aiken, South Carolina, and across the northern two-thirds of the Florida peninsula and adjacent northeastern Georgia. Working westward it is found in Walton and Okaloosa counties, Florida, then again from Mobile Bay, Alabama, to southeastern Louisiana.

Habitat: The pine woods snake is a resident species in pinelands that are well on the damp side of dry. Where conditions are shaded, damp, and humid, look for this snake beneath any manner of surface debris—both natural and human generated.

Size: This tiny snake is often only the length and diameter of a pencil, with a normal adult size of 8–12 inches and a record of just under 16 inches. Hatchlings are about 5 inches long.

Identifying features: The dorsal color is yellowish brown to reddish brown, darkest vertebrally. The chin is white. The venter is off-white, pale yellow, or pale green, and unmarked. There is a dark eyestripe, and the upper labials are light in color (but are seldom even vaguely yellowish) and bear a variable amount of dark flecking. The smooth scales are in 17 rows and the anal plate is divided.

52. Pine Woods Snake

Similar species: Earth snakes, brown snakes, and red-bellied snakes have keeled scales. Crowned snakes have black heads.

Rat Snakes, Kingsnakes, Bullsnakes, and relatives, subfamily Lampropeltinae

We have chosen to use the Lampropeltinae as a subfamily of the Colubridae, even though the designation currently seems to have fallen into disfavor with systematists. It is a convenient repository for this grouping of snakes that seem more closely related to each other than to the snakes in other subfamilies.

There are eight genera of lampropeltine snakes in the eastern and central United States. Besides the well-known rat, pine, and king snakes, there are the glossy, long-nosed, scarlet, and short-tailed snakes. Most of these snakes are powerful constrictors that kill their prey before eating it. The short-tailed snake is a weak constrictor that specializes in small snakes as prey, merely restraining them from coiling into unmanageable positions after seizing. The reptile-egg eating scarlet snakes have no need to constrict, and seldom even coil tightly.

The lampropeltine snakes are the favorites of herpetoculturists, and tens of thousands of these snakes are bred in captivity every year. Aberrant (designer) colors and patterns have been developed in many species. Breeding habits in the wild are more poorly known. Despite their very different appearance, the kingsnakes, pine snakes, and rat snakes have been hybridized by hobbyist-breeders and, in most cases, the offspring produced have proven reproductively viable. Captive-bred snakes of any kind should never be released into the wild.

Representatives of this subfamily occur widely over North America, Europe, and Asia. A few types are present in the Neotropics.

All lampropeltines are nonvenomous.

Finding lampropeltids

The first corn snake I (RDB) ever saw was beneath a piece of tin near an old barn in the low country of southeastern South Carolina. I was maybe 15 years old and had dreamed of finding just such a creature. The phase that I found was called the "Okeetee phase" (the name being taken from the Okeetee Hunt Club of the area) and noted for the richness of its red on orange coloration. Eventually I found other color phases from other areas, but few were as beautiful as that first example.

Today, because of the efforts of hobbyists, captive-bred corn snakes are available in dozens of designer colors and patterns. They can now be seen in almost every pet store in North America, Europe, and some areas of Asia. They are the shining, multi-hued stars of herpetocultural expos now held in venues as diverse as fairgrounds and prestigious hotels—venues where only a few years ago these creatures would have been anything but welcome.

But not all of the snakes of this family are as common and well understood as the corn snake. For example, although the scarlet snake is common, the life history of this little snake, and especially of the Texas race, remains enigmatic. And the very specialized short-tailed snake is not only of local distribution but also thought to be very rare. Despite having lived in short-tailed snake habitat for most of my adult life, I have found only three of these snakes myself and seen only four additional examples.

So, from the common to the rare, and from the brightly colored to the plain, the rat snakes, kingsnakes, and relatives have become the North American flagship snakes, with a popularity equaled by no other snake species worldwide.

Glossy Snakes, genus *Arizona*

This is a pretty snake, represented in the central United States by a single species and numerous subspecies. None are brightly colored but all are attractive. In appearance the glossy snake is most reminiscent of the bull and gopher snakes, but the former have smooth scales and a less accentuated rostral (nosetip) scale. Glossy snakes have only 2 prefrontal scales. Although some examples of the glossy snake are strongly patterned, many have a hazily defined pattern of faded blotches. From this they derive the often-heard colloquial name of faded snake. Striped, rather than blotched, examples have been found.

The average number of dorsal blotches and the number of ventral scales are two of the defining characteristics among subspecies, but these can be both variable and difficult to determine unless the snake is in hand. Ventral scale counts, for example, vary by sex, but counting them on an unrestrained living snake is very difficult. A glossy snake found where ranges abut and intergradation occurs may be impossible to accurately assign to a subspecies.

The relative harshness of the aridland habitats of these snakes probably restricts females to a single clutch of eggs annually. Between 5 and 15 eggs (rarely to 23) have been recorded. Depending on climatic conditions, incubation takes 60–90 days.

Usually inactive during the hours of daylight, the glossy snakes emerge from hiding and begin to prowl soon after dark on warm and hot evenings. They may remain active until late into the evening and are often encountered as they cross desert and agricultural roads. During the day glossy snakes seek shelter in the vacated burrows of small mammals, burrows of their own making, or amid or beneath rocks. If confronted while active, a glossy snake may assume a striking S shape and vibrate its tail energetically. This produces the whirring sound so often associated with rattlesnakes if the snake is in dry grasses or leaves.

The glossy snakes are constrictors that prey heavily on lizards and small mammals; they also eat nestling birds.

53. Kansas Glossy Snake

Arizona elegans elegans

Disposition: These snakes are usually reluctant to bite.

Abundance/Range: This is a common snake throughout most of its range, which extends southward from eastern Colorado and extreme southwestern Nebraska, through western Texas, and well southward into eastern Mexico.

Habitat: This is a snake of rocky semi-aridlands, sandy fields, pasturelands, and sparsely wooded canyons.

53. Kansas Glossy Snakes

Size: Although they may rarely reach 48–52 inches in overall length, most adult glossy snakes are much smaller. The more typical size is 30–42 inches. Hatchlings are about 10 inches in length.

Identifying features: All subspecies of the glossy snake are normally blotched with large dark dorsal saddles and smaller dark lateral spots against a variably light background. The Kansas glossy snake usually has 50 or fewer dark vertebral blotches. The average blotch count is 53. The blotches are wide and usually surrounded by thin, darker borders. Striped individuals have been documented. The colors—buff, tan, or light brown—usually blend well with the sand and rocks among which the snake is found. The lightest areas of ground color sepa-

rate the dark dorsal blotches vertebrally. Glossy snakes may darken with age, further obscuring the dark blotches. The white belly is unmarked. The head is narrow. The body scales are smooth. The anal plate is single. Table 1 compares identifying features of the three subspecies of glossy snakes.

Similar species: The bullsnake and gopher snakes have strongly keeled body scales and 4 prefrontal scales. The rat snakes have weakly keeled dorsal scales. Kingsnakes have the dorsal scales peppered with light spots and strongly delineated light CROSS BANDS.

Table 1. Identification features of subspecies of Glossy Snakes, Arizona elegans

Characters	arenicola	elegans	philipi
Common name	Kansas	Texas	Painted Desert
29 or more scale rows at midbody	x	x	
27 or fewer scales at midbody			x
Males with 211 or fewer ventral scales		x	
Females with 220 or fewer ventral scales		x	
Males with 212 or more ventral scales	x		
Females with 221 or more ventral scales	x		
41–58 vertebral blotches	x		
40–69 vertebral blotches		x	
Usually 62 or more vertebral blotches			x

54. Texas Glossy Snake

Additional subspecies

54. The Texas Glossy Snake, *Arizona elegans arenicola*, is the southeasternmost representative of the genus. See Table 1 for characteristics, but depend mostly on range for identification; it is found in central and southern Texas. It, too, is a wide-blotched species, but the blotches may or may not have dark borders. Average blotch count is 49. This race tends to be of pallid coloration.

55. The Painted Desert Glossy Snake, *Arizona elegans philipi*, is found in extreme western Texas, New Mexico, southeastern and northeastern Arizona, and extreme southern Utah. For identification use both Table 1 and the range map. The ground color of the Painted Desert glossy snake is a pretty buff. The saddles are a dark-edged, darker brown and of intermediate width.

55. Painted Desert Glossy Snake

Scarlet Snakes, genus *Cemophora*

This genus contains only a single species but three subspecies. All are persistent burrowers and of beautiful color and pattern. Even when severely provoked scarlet snakes rarely bite. The scarlet snake often surfaces on warm, humid nights or may be active above ground when its burrow floods during heavy rains. The scarlet snake is clad in bands (bands do not encircle the body) of red, black, and white or yellow. Bands of black separate the two caution colors of yellow and red. The head is narrow and only slightly distinct from the neck; the pointed snout is red. The belly is an unbanded off-white to yellow. The smooth scales are arranged in 19 rows and the anal plate is undivided.

Like all other North American members of this subfamily, the Florida scarlet snake is an oviparous species. Clutches are small, usually numbering fewer than 8 eggs. It is probable that only a single clutch is laid each year in very late spring or early summer. Incubation takes 55–70 days.

The scarlet snake is generally not seen above ground until well after darkness has fallen. It occasionally may be found during the day by turning ground-surface debris. It may also be induced to diurnal activity by the lowered barometric pressure that accompanies heavy rains.

Contrary to anecdotal accounts of the scarlet snake accepting lizards, smaller snakes, and even newly born mice as prey, none of the many specimens we have observed in captivity have voluntarily done so. Rather, all held out until lizard or small snake eggs or broken turtle eggs were offered. The snakes drink the contents. Captives occasionally accept pre-killed nestling mice that have been dipped in egg.

Despite its affinities to the kingsnakes, pine snakes, and rat snakes—powerful constrictors all—the scarlet snake is not a noted constrictor.

56. Florida Scarlet Snake

Cemophora coccinea coccinea

Disposition: This snake seldom may be induced to bite.

Size: Normally 14–20 inches long when adult, occasional examples may attain a length of 30 inches. The hatchlings are just under 6 inches in length.

Abundance/Range: This is another of the many eastern snakes that are quite common, but so secretive that they are seldom seen. The Florida scarlet snake inhabits areas of loose soils on the southern two-thirds of the Florida peninsula.

Habitat: The Florida scarlet snake may be seen, sometimes in numbers, after

56. Florida Scarlet Snake

Florida
northern
Texas

darkness has fallen, crawling on the sandy woodland floor, or crossing sand-country roadways. Look for it in sandy scrub as well as in elevated hammocks, in pinelands, on dry prairies, even beneath debris on weedy road shoulders.

Identifying Features: This is a brilliantly banded snake with a pure white belly. The pointed snout is bright red, which is usually the predominant body color as well. The yellow (or white) and black bands are narrower than the red. The black and yellow bands may actually be true bands (or may be little more than saddles) but are never rings involving the belly scales. The color sequence is red-black-yellow-black-red. The pattern may be disrupted, one color encroaching on the next, on very old snakes. This subspecies normally has 7 supralabial (upper lip) scales, 19 rows of smooth scales, and an undivided anal plate.

Similar species: The scarlet kingsnake and the coral snake are clad in *rings* (not bands) of red, black, and yellow. Additionally, the coral snake has the two caution colors, red and yellow, touching, not separated by black. The northern scarlet snake usually has only 6 upper labial scales.

Comments: The scarlet snakes and several races of milksnakes are often confused with the dangerously venomous coral snake. In most cases the snakes are easily differentiated by band arrangement.

The nonvenomous scarlet snakes, scarlet kingsnake, and milksnakes have the two caution colors, red and yellow, *separated by a band of black.* Coral snakes have the caution colors touching (except on the Florida Keys, where many examples of the coral snake lack much or all of the yellow). Also see accounts 83, 86, 89, 90, 92, 93, 185, and 186.

Additional subspecies

57. The Northern Scarlet Snake, *Cemophora coccinea copei*, is very similar to its southern relative, but usually has only 6 supralabial (upper lip) scales. This snake inhabits areas of loose soils on the northern one-third of the Florida peninsula and from there north in the coastal plains and piedmont to the Pine Barrens of New Jersey and eastern Texas.

57. Northern Scarlet Snake

58. Texas Scarlet Snake. Photo by Paul Freed

58. The Texas Scarlet Snake, *Cemophora coccinea lineri*, is the least well known of the trio of subspecies. It is found from just south of Galveston Bay to the area around Edinburgh, Hidalgo County, in South Texas. This race differs in being clad in paler red and in having more than 183 ventral scales. The red is also more extensive than in the other races, and is not margined with black at its lower extremes. Range alone will identify this seldom-seen snake.

Rat Snakes, genera *Bogertophis* and *Elaphe*

Although *Pantherophis* is the newly proposed generic name for the snakes of the genus *Elaphe*, most researchers suggest that, until further studies are completed, the use of *Elaphe* should be retained.

The genus *Bogertophis* is represented by two species, only one of which (the Trans-Pecos rat snake) occurs in our area. The genus *Elaphe* is represented by four (some researchers feel there are six or more) species. These are the various races of the corn snake, two subspecies of the fox snake, Baird's rat snake, and the several races of the black rat snake.

As a group, when adult, these rat snakes have weakly keeled dorsal and dorsolateral scales and unkeeled lateral scales. The juveniles of all lack scale keels. The number of scale rows varies both by species and by individual. The anal plate is divided. All American species in this genus are oviparous.

In cross-section, these snakes are rounded on top and have weakly convex sides and a flattened venter. Rat snakes are agile climbers that can ascend virtually straight up a tree with only moderately rough bark and can even ascend smooth-barked trees (though with more difficulty).

Juvenile rat snakes usually feed on small lizards and treefrogs, while the adults of most species consume rodents, rabbits, bats, and some birds. The Trans-Pecos rat snake will accept lizards throughout its life.

Most rat snakes kill or immobilize their prey animals by constriction. Some, however, fail to constrict small prey. This is especially true of the Trans-Pecos rat snake and the fox snakes that often pursue rodents into their burrows, where constriction would be physically impossible. Rather than constrict, these snakes immobilize their prey by pressing them against the side of the burrow with a body curve.

If carelessly handled, most wild rat snakes will bite, and some strike savagely and repeatedly. Many quickly become accustomed to handling and become very tractable. All are nonvenomous.

Desert Rat Snakes, genus *Bogertophis*

These nocturnal snakes are slender, nervous, big-eyed serpents that are found in desert and rocky savanna and riparian scrubland. Some researchers consider the two snakes in this genus more closely allied to the bullsnakes and gopher snakes than to the rat snake genus *Elaphe*, from which *Bogertophis* was only recently separated. There are only two species in this genus. One, the Baja rat snake, occurs in extreme southern California and over much of the Baja Peninsula. The second, the Trans-Pecos rat snake, is found in western Texas, south central New Mexico, and north central Mexico.

The head is broad, somewhat flattened, and distinct from the slender neck. The body scales are weakly keeled and arranged in 31–35 rows, and the anal plate is divided.

The Trans-Pecos rat snake breeds much later in the year than many other snakes. Adult males are commonly found crossing desert roadways in May, June, and July. It is thought that many of these are following the pheromone trails of receptive females. Breeding males are known to fight savagely with each other. Egg deposition occurs in June, July, and August. Incubation is variable, taking 62–75 (rarely more than 100) days.

Most Trans-Pecos rat snakes will allow gentle handling without showing a defensive reaction.

59. Trans-Pecos Rat Snake

Bogertophis subocularis subocularis

Disposition: This snake usually does not attempt to bite when encountered in the wild. However, it is fully capable of doing so, and some may strike repeatedly.
Abundance/Range: Long thought to be a rare snake, it is now known that the Trans-Pecos rat snake is retiring and nocturnal, but quite common. This snake

59. Trans-Pecos Rat Snakes

is a resident of the Chihuahuan Desert. It may be found southward from south-eastern New Mexico, through western Texas, to central Durango, Mexico.

Habitat: This is a snake of boulder fields, rocky hillsides, escarpments, road cuts, creviced cliff faces, and other similar habitats. It is particularly common in rocky riparian areas that support tangled growths of desert scrub such as ocotillo, agave, creosote bush, shin oak, and cholla.

Size: Occasional specimens of the Trans-Pecos rat snake may measure 66 inch-es in length. Most adults, however, are considerably smaller, measuring 36–52 inches. Except when in their largest sizes, when they tend to become heavy bod-ied, the Trans-Pecos rat snake is a slender serpent. Hatchlings measure 11½–14 inches in length.

Identifying features: This big-eyed rat snake is restricted in distribution to the

Chihuahuan desert of the American Southwest and northern Mexico. In its most characteristic color and pattern, it is one of the most distinctive snakes of the Chihuahuan desert. Two lesser known colors also occur.

Typically, 21–30 black H-shaped markings are in strong contrast to a ground color ranging from straw yellow to olive yellow to tan. The venter is unmarked and of an opalescent off-white to pale tannish yellow in coloration.

In the Franklin Mountains of West Texas the Trans-Pecos rat snake has the typical black H-shaped markings, but a gray ground color. The unmarked venter is a pale silvery gray.

In both of the above phases, the neck is marked by 2 wide, parallel black stripes. On the posterior neck the crossbars of the Hs begin. The H pattern weakens posteriorly, and the tail is banded. White markings may show between the scales (interstitially), especially where the crossbars of the H join the dorsolateral stripes. A series of lateral blotches, often poorly defined, are usually visible on each side.

In the Christmas Mountains (and surrounding areas), in the Big Bend region of Texas, a very small percentage of the Trans-Pecos rat snakes are of a brighter yellow, atypically marked phase. These blonde-phase rat snakes lack neck bars and H-shaped dorsal markings and instead have simple, often poorly defined, darker dorsal saddles. If lateral blotches are present on this phase, they are vague at best.

Hatchlings are proportionately slender, have huge eyes, and appear translucent when viewed ventrally.

In all phases, the dorsal surface of the head is unpatterned. There is a row of small scales (the suboculars) between the eye and the upper labial scales. The dorsal scale rows are in 31–35 rows. There is weak keeling on the dorsal and dorsolateral scale rows, but the lateral scale rows are smooth. The anal plate is divided.

Similar species: There are no other snakes with the H pattern so typical of the normal and gray phases of the Trans-Pecos rat snake. The large eyes and subocular scales will differentiate Trans-Pecos rat snakes of the blonde phase from glossy snakes or other rat snakes.

Typical American Rat Snakes, genus *Elaphe*

These are among the most popular snakes with herpetoculturists, persons who breed reptiles and amphibians in captivity. The rat snakes in this genus have weakly keeled dorsal and dorsolateral scales and unkeeled lateral scales. The juveniles of all lack scale keels. The number of scale rows varies. The corn snakes have the scales arranged in 27 or 29 rows, the fox snakes in either 25 or 27 rows, and the black rat snake and its relatives in 25–33 rows. The anal plate is divided.

Because the various rat snakes are often found in the vicinity of farms and barnyards, especially near piles of debris or unused buildings, farmers often refer to them as chicken snakes. While there is no denying that a rat snake will eat an occasional baby chick (and even more rarely, an egg or two), it is usually the proliferation of rodents, the preferred prey of these powerful constrictors, that draws them to farmyards.

The Baird's rat snake and some subspecies of the black rat snake undergo marked ontogenetic (age-related) changes. At hatching, the juveniles of all are strongly blotched or crossbarred. With age and growth the Baird's, the yellow, and the Everglades rat snakes lose their juvenile pattern, becoming, instead, strongly striped. Adults of the black, gray, and Texas rat snakes, the corn snake, and the fox snakes are larger renditions of the blotched babies.

Juvenile rat snakes usually feed on small lizards and treefrogs, while the adults of most consume rodents, rabbits, and some birds.

Based primarily on molecular data, an overhaul of the long-standing taxonomy for the *Elaphe obsoleta* group has recently been suggested. Four clades (phylogenetic groups) have been named and the use of subspecific names discontinued. The newly proposed names are *Elaphe alleghaniensis* (eastern rat snake) in the East, *E. spiloides* (central rat snake), next west, *E. obsoleta* (western rat snake) from west of the Mississippi, and *E. bairdi* (Baird's rat snake) from central Texas and northern Mexico. As defined, the yellow and the Everglades rat snakes would be grouped together as *E. alleghaniensis*, while the gray rat snake would be divided between *E. alleghaniensis* (northwestern peninsula and eastern panhandle of Florida) and *E. spiloides* (western panhandle of Florida northward).

It has recently been proposed that the generic name *Elaphe* be replaced by *Pantherophis* and the Kisatchie corn snake designated *P. slowinskii*.

At breeding time the males of some rat snakes engage in stylized combat activities. Males approach one another, entwine bodies, raise the entwined anteriors, and try to topple and suppress each other. This ritualized effort to assert dominance can continue for some time, and is often followed by the dominant male successfully breeding the female.

When food is sufficiently abundant that female rat snakes in the Deep South are able to maintain a good body weight, two clutches of eggs may be laid annually. Northern populations lay only a single clutch each year. The eggs may number 5–30 per clutch with more usually produced in the first clutch than in the second. Incubation takes 55–70 days. Late hatchlings in northern locales (and especially in the Great Lakes region, where weather conditions are notoriously uncertain) may hibernate before eating the first time.

Corn snakes, fox snakes, and Baird's rat snakes tend to be more terrestrial than the black rat snake and its relatives. Baird's rat snakes may often be seen basking in the afternoon sunshine, draped across or upon the limbs of a tree or atop a dense shrub. They also select crumbling stone walls and deserted buildings as hiding areas. Rocky jumbles and fissured canyon faces are typical habitats of the Baird's rat snake. Rat snakes also thermoregulate beneath surface debris such as stored roofing tin or fallen buildings. These snakes prowl extensively late into the afternoon and, temperatures permitting, well into the hours of darkness. They may be especially active during periods of low barometric pressure and during the spring breeding season. Males then diligently follow the pheromone trails left by receptive females. Rat snakes often bite defensively.

60. Baird's Rat Snake

Elaphe bairdi

Disposition: This snake is apt to bite if startled or if handled carelessly.
Size: Although the record size is 62 inches, most Baird's rat snakes are 36–54 inches in length. Hatchlings are about 11 inches long.
Abundance/Range: Baird's rat snake was once thought to be an uncommon species. However, snake hunters driving the roads of West Texas after dark have learned that this snake is more secretive than rare.
Habitat: The Baird's rat snake is often found in rocky montane canyons, on rocky, sparsely vegetated desert hillsides, in creviced road cuts, on aridland roads, in rocky areas near desert waterholes or rivers, and in other similar habitats.
Identifying features: Although the colors of the adult Baird's rat snake are certainly not brilliant, the overall appearance is of a remarkably attractive snake. Adults of this snake are yellowish brown to grayish brown with an opalescent overlay on the scales. The color tends to be richer posteriorly. Four darker lines, 2 dorsolaterally and 2 laterally, are visible and often most distinct anteriorly.

60. Baird's Rat Snake

Vestiges of dark blotches may remain visible through early adulthood. The belly is yellowish anteriorly, darkening to orangish posteriorly. Dark smudges or blotches are often present. The top of the head is slate gray and the face may be gray to yellow. The juveniles are grayish, have a dark bar across the top of the head connecting the eyes, and have 48 or more narrow cross bands that do not have light centers.

Similar species: Patch-nosed snakes have a wraparound rostral (nosetip) scale and are very slender. Adults of the Texas rat snake are strongly blotched.

Comments: This snake was long considered the westernmost subspecies of the black rat snake. Although noted similarities in appearance exist, it is now considered a full species.

61. Corn Snake

Elaphe guttata guttata

Disposition: Although it is usually of reasonably good disposition, a corn snake may occasionally strike and bite when molested. This is especially true, as with most other snakes, when it is preparing to shed its skin.

Abundance/Range: This is an abundant snake throughout much of its very extensive range. However, in peripheral areas of its range it may be uncommon. It is protected by some states. This snake ranges from the Pine Barrens of New Jersey to western Tennessee and western Louisiana and to the southernmost tip of the Florida Keys.

Habitat: This is a snake of deciduous woodland edges, mixed woodlands, and

61a. Corn Snake, red morph

61b. Corn Snake, gray morph

61c. Corn Snake, anerythristic morph

corn
Great Plains
southwestern

pinelands. It occurs in meadows, pastures, and prairies. In Florida it can be abundant in roadside windbreaks of Australian pines or Brazilian pepper. It survives, seemingly with little effort, in agricultural areas, suburban rock piles, city yards, and hedgerows. It particularly favors old outbuildings, stone walls, trash heaps, hay and feed storage areas, or any other habitat that induces, or channels the movement of, healthy rodent populations. Corn snakes are agile climbers and can occasionally be found rather high in trees and in tree hollows (where they may hibernate), but are less apt to climb than certain other rat snakes. They also may be found in rodent burrows and caves or caverns (but do not seem to go far into the caves). They may be seen in the rafters or crawl spaces of dwellings and barns, or coiled quietly in low shrubs or among the stored pots in greenhouse nurseries. In other words, the corn snake is a survivor species and a habitat generalist.

Size: This snake is of variable size. In South Florida where its diet often consists largely of treefrogs and lizards, the corn snake may be mature at 26–30 inches. Elsewhere, where its diet is more varied and strongly oriented toward rodents, a length of more than 48 inches is often attained. The largest corn snakes can exceed 6 feet in length. Hatchlings are about 8½–12 inches in length.

Identifying features: Although corn snakes are usually readily recognizable throughout their range, they do vary widely in dorsal color. Most have a well-defined dark spearpoint marking on the top of the head. No matter the color phase, there are strongly defined dorsal saddles and a row of smaller lateral spots on each side. Some old individuals, especially when preparing to shed, may show 4 indistinct, dark longitudinal stripes.

The corn snakes of the Atlantic coast are usually tan or red with black-edged, deeper red blotches. The species seems particularly brightly colored from eastern North Carolina southward to northeastern Florida. Specimens from the more westerly portion of the range often have an olive wash over both ground color and blotches, with less contrast between blotches and ground color.

Corn snakes are particularly variable in Florida. A fair proportion of corn snakes from northeastern Florida have much reduced black and a rosy wash over the entire dorsum. A few very red (erythristic) corn snakes lacking virtually all of the black (both dorsally and ventrally) have been found in this population.

Interior southwestern Florida is home to a significant population of corn snakes that lack virtually all red pigment. Areas of yellow are often visible on the sides of the neck. Except for the yellow highlights, these anerythristic ("lacking red") corn snakes have a grayish ground color with dark-edged brown dorsal saddles. The dorsal saddles are often outlined in black and the belly is normally black and white checkered.

The corn snakes of central western Florida often have a reduced amount of black pigment both dorsally and ventrally; those of South Florida are typified by a silvery gray dorsum with black-edged, deep red saddles.

The corn snakes of the Florida Keys (61d) are equally variable. Despite having once been called rosy rat snakes, many are anything but rosy. While the saddles of all vary from rosy to dark red, the ground color may be olive, light

61d. Keys Corn Snake, olive phase

61e. Corn Snake (western Louisiana)

rose red, silver, or, more rarely, brownish. There is a tendency for all black pigment to be reduced, and the spearpoint on the head is often indistinct.

The corn snakes of western Louisiana (the Kisatchie region) are of particularly dark coloration (see photo 61e) and intermediate in appearance between the corn snake and the Great Plains rat snake.

The belly scales of most corn snakes are strongly patterned with black checkers against a ground color of white. Wide black stripes are usually present beneath the tail. Examples from populations in northeastern Florida and the Lower Florida Keys may lack most black on the belly and tail.

The weakly keeled scales of all corn snakes are in 27 (normal) or 29 rows and the anal plate is divided. See Table 2 for identifying features of the subspecies of corn and rat snakes.

Similar species: Some specimens of the red phase of the mole kingsnake and prairie kingsnake may look superficially like a corn snake. However, the kingsnakes have a narrow head, no keels on any scales, and an undivided anal plate.

Table 2. Identification features of subspecies of Corn and Rat Snakes, Elaphe guttata

Characters	guttata	emoryi	meahllmorum
Common name	Corn	Great Plains Rat	Southwestern Rat
Dorsal blotches 44–59 (average 51)	x		
Dorsal blotches 57–81 (average 67)		x	
Dorsal blotches 39–67 (average 55)			x
Strongly checkered belly	x	x	
Weakly or noncheckered belly			x
2 well-defined subcaudal stripes	x	x	
Weakly defined subcaudal stripes or spots			x

62. Great Plains Rat Snake

Additional subspecies

62. The Great Plains Rat Snake, *Elaphe guttata emoryi*, is the race of corn snakes found west of the Mississippi River and north of South Texas. It is very much like a dark corn snake in appearance. The ground color is tan to olive tan, occasionally with a vague blush of orange dorsally. The dorsal saddles may be a darker olive tan, olive brown, dark brown, or almost black. The dark saddles often have lighter centers and are edged on both their fore and aft edges (sometimes on the lower edges as well) with black pigment. The belly is very strongly marked with black checkers against a white ground color. The subcaudal scales bear 2 black stripes. This snake is somewhat smaller than its eastern relative, seldom exceeding 54 inches in length. The record length is 60¼ inches.

63. The Southwestern Rat Snake, *Elaphe guttata meahllmorum*, is the South Texas representative of the corn snake group. It averages fewer dorsal blotches than the Great Plains rat snake, and is usually a warmer tan or brown on both ground and blotch color. The belly is often unpatterned. See Table 2 and the range map for identification assistance.

Comment: The herpetocultural community has bred corn snakes extensively for more than three decades. During this time, more than forty "designer" colors and patterns have been developed. These vary from simple albinistic specimens to caramel colors, butter yellows, and piebalds, and from normal blotches to saddles and even stripes. Since snakes are legendary for their Houdini-like escape acts, it is feasible that you could meet a roaming corn snake of virtually any color, either in or far out of the snake's normal range. These may be difficult

63. Southwestern Rat Snake

to identify, but in most cases the telltale spearpoint will be visible on the top of the head.

64. Black Rat Snake

Elaphe obsoleta obsoleta

Disposition: Individuals of this large snake vary tremendously in temperament. While many specimens are easy going, others will strike savagely and repeatedly. Their jaws are strong and their teeth can produce lacerations that may bleed freely.

64. Black Rat Snake

black
Texas
yellow
Everglades
gray
Gulf Hammock

Abundance/Range: This is a commonly encountered snake over much of the northeastern United States and southern Ontario. It ranges southward from central western Vermont and southeastern Minnesota to central South Carolina, northern Louisiana, and central Kansas.

Habitat: This is a snake of deciduous and mixed woodlands and of mountain and lowland clearings. It is often found in the proximity of human habitations where it particularly favors old outbuildings. Boulder-strewn hillsides, escarpments, stone walls, and trash heaps are also utilized by this snake. It often lies next to rodent trails.

Size: This snake and its relatives are among the largest serpents of the eastern United States. Specimens measuring 54–66 inches are considered average-sized adults. Individuals of more than 72 inches in length are not particularly uncommon, and the greatest recorded size is 102 inches! Hatchlings are 11–15 inches long.

Identifying features: Adults of this powerful constrictor vary from grayish brown to brown, to almost jet black. White, yellow, and/or orange red may be present interstitially (on the skin between the scales) and often becomes visible if the snake is distended with food or is heavily gravid. The belly is predominantly white but bears irregular small checkers of dark pigment. The chin and throat are white. At the periphery of its range, where it intergrades with other subspecies, confusing suites of colors may be encountered (see comments about the greenish rat snake, an intergrade between the black rat snake and the yellow rat snake, in the Additional Subspecies and Intergrades section that follows). Juvenile black rat snakes have a gray ground color, black dorsal blotches, and black lateral spots between the dorsal blotches. They darken more with every shedding of the skin, soon assuming the adult coloration. A dark interorbital line is present. This continues posterior to the eye and stops at the mouth line. If the underside of the tail is striped, it is only weakly so. The scales are in 27 rows and those in the dorsal rows are weakly keeled. The anal plate is divided.

Similar snakes: Black racers have smooth, satiny black scales and flee rapidly when approached. Black-phase hog-nosed snakes have an upturned rostral (nosetip) scale. Water snakes of all kinds have very strongly keeled body scales.

Comments: The colors and patterns of some examples of the black rat snake, the gray rat snake, and the Texas rat snake are confusingly similar. All races of this snake readily intergrade with each other along the peripheries of their ranges. This produces offspring that are of intermediate appearance.

Subspecies and intergrades

65. The Texas Rat Snake, *Elaphe obsoleta lindheimeri*, is the most southwestern representative of this species. It ranges throughout most of the eastern half of Texas as well as southern Louisiana. It occupies a wide variety of habitats, from

65. Texas Rat Snake

the bayous of Louisiana to woodlands and semi-arid canyons in Texas. The Texas rat snake (occasionally referred to as Lindheimer's rat snake) starts life as a very dark-colored hatchling and doesn't change much during growth. Adults have saddles of black, bluish black, or dark brown. These contrast strongly against the ground color of lighter brown, tan, or yellowish.

Intergrades between the Texas rat snake and the black rat snake occur over much of southern and western Oklahoma.

66. The Yellow Rat Snake, *Elaphe obsoleta quadrivittata,* is a beautiful subspecies. Adults of this powerful constrictor vary from gray or greenish gray with 4 wide and well-defined dark stripes in the coastal Carolinas, to a rich yellow with variably dark stripes in South Florida. (The dark examples of the Carolina coastal plain are often intergrades of yellow and black rat snakes. These are referred to as "greenish rat snakes.") South of Lake Okeechobee, an area where the influence of the now uncommon Everglades rat snake still remains, the ground color may near orange and the stripes may be poorly defined. The tongue is dark (with a red base in the south), the eyes are yellow to orange, and the chin is largely white. The belly is pale yellow. The 27 rows of scales are weakly keeled and the anal plate is divided. The yellow rat snake is found in both wet and dry habitats from Pamlico Sound, North Carolina, to the Florida Keys.

A color variant of the yellow rat snake (66b) is found on Key Largo, southward at least to Marathon, and immediately north of Card Sound on the Florida mainland. It is uncommon on the Keys, occurring in the coastal mangrove

66a. Yellow Rat Snake

66b. Yellow Rat Snake, Key Largo form

strand as well as in the interior hardwood hammocks. It was once called "Deckert's rat snake" and afforded the now invalid designation of *E. o. deckerti*. This variant has a rusty orange ground color (some specimens have a silvery sheen overlaying the orange). Besides striping, it also has variably distinct dorsal saddles. Juveniles are patterned similar to, but more precisely than the adults and tend to have both ground color and saddles more brown than orange.

Juveniles of the yellow rat snake are as variable in ground color as the adults. Northern specimens are dark gray with black dorsal blotches. Southernmost specimens are light gray with dark gray dorsal blotches. Juveniles of the Keys variant (see next account) usually have an orangish wash to the light areas and brown dorsal blotches.

67. The Everglades Rat Snake, *Elaphe obsoleta rossalleni*, is the most brightly colored rat snake of North America. Pure-strain adults of the Everglades rat snake have an orange body color, deep orange eyes, an orange chin, and an all-red tongue. The stripes are weakly defined. The belly is paler than the back, but orange nonetheless. Juveniles are buffy orange and have well defined brownish dorsal saddles. Once common, with the draining of the Everglades this magnificent snake has been overwhelmed by the encroachment of the yellow rat snake, a race preferring the drier habitats. Intergrades between the two races are now the norm over most of the Everglades rat snake's original range.

This is a snake of those few flooded sawgrass/marl prairies that remain. Look for it in isolated hammocks, along canals, and in 'Glades-edge windbreaks. Stone piles, limestone outcroppings, heaps of vegetation, and trash heaps are

67. Everglades Rat Snake

also favored habitats. These rat snakes are adept at climbing and often found high in trees. Juveniles seem to climb less than the adults.

This snake is restricted in distribution to the Kissimmee Prairie and the Everglades of interior southern Florida.

68. The Gray Rat Snake, *Elaphe obsoleta spiloides*, is more readily recognized by its colloquial name of white oak snake than by its accepted common name of gray rat snake. Where the color of this impressive snake is not compromised by genetic input of other rat snake races, the ground color is just that—gray. The dorsal blotches are dark gray and often have a lighter center. There are often dark vertical markings along the edges of each upper labial scale and a dark diagonal line from rear of the eye to the corner of the mouth. The belly may be a lighter gray than the back and usually has a double row of small, darker

68. Gray Rat Snake

markings. Some white may be visible interstitially, especially when the snake is distended with food or heavily gravid. Hatchlings have a gray ground coloration and dark brown blotches.

Gray rat snakes are agile climbers and during hot weather may remain draped over a limb for days on end. This is a snake of deciduous and mixed woodlands as well as of pinelands. It may be found in clearings, near human habitations, in infrequently used outbuildings, stone walls, trash heaps, barns, or wherever else there may be a concentration of rodents. The range of the gray extends westward and northward from the Florida panhandle to western Mississippi and southern Illinois.

69. The variant known as the Gulf Hammock Rat Snake, a snake of the varied woodlands of Florida's Gulf Hammock region, is an intergrade between the gray and the yellow rat snakes. It was once given the scientific name of *E. o. williamsi*. This snake looks much like a light-colored, often olive yellow, gray rat snake (or, conversely, a dull-colored yellow rat snake), but has dark stripes as well as dark dorsal blotches. The belly is tannish gray and often bears a double row of dark spots. There are often dark vertical markings along the edges of each upper labial scale and a diagonal postocular line running from the back of the eye to the corner of the mouth. Some white may be visible interstitially, especially when the snake is distended with food or heavily gravid. Hatchlings have a gray ground coloration and dark brown blotches.

Although the range of this intergrade form is primarily in the vicinity of Florida's Gulf Hammock, it extends from Crystal River to Steinhatchee, then northeastward to the Osceola National Forest.

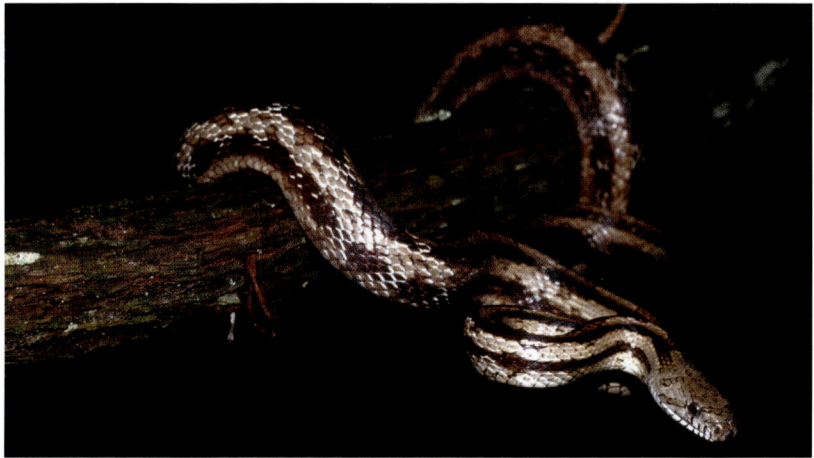

69. Gulf Hammock Rat Snake

70. Western Fox Snake

Elaphe vulpina vulpina

Disposition: The western fox snake is usually a quiet and inoffensive snake but some individuals may bite if molested.

Size: Adults occasionally reach about 60 inches (more normal is 36–48 inches) and hatchlings are about 11 inches in length.

Abundance/Range: This snake remains fairly common throughout much of its range, westward from the western shore of Lake Michigan and the southern shore of Lake Superior to eastern South Dakota and Nebraska.

Habitat: The western fox snake occurs in open woodlands, prairies, hayfields, and associated areas.

Identifying features: The ground coloration of the western fox snake may be buff, yellowish, or off-white. The 32–52 (average 41) well-defined dorsal blotches are often of irregular outline and dark brown to black in color. The first blotch behind the head is often elongate, of an H or U shape, with the arms directed forward. Lateral blotches alternate between blotches on the dorsum. The first lateral blotch is noticeably elongate. Both the head and tail may be brighter (or darker) than the ground color. Typical head/tail colors are olive, brownish, buff, copper, or a rather bright orange. The cheeks may be brighter than the crown. Young fox snakes have 2 dark bars across the crown, one interorbitally and the other paralleling the posterior edge of the prefrontals. Another dark bar extends from each eye to the back of the mouth. The head markings often fade, sometimes to invisibility, with advancing age. Keeling is strongest on the several most dorsal rows of scales. The anal plate is divided. The ground coloration of the black-checkered belly is cream to yellow.

70. Western Fox Snake

western
eastern

Similar snakes: The visible identifying characteristics of the two races of fox snakes overlap widely but the ranges of the two snakes are widely separated. The eastern fox snake normally has fewer (average of 32) and larger dorsal blotches. To separate juveniles of the western fox snake from juvenile black rat snakes it may be necessary to count the number of belly scales. Black rat snakes have more than 220, western fox snakes have fewer than 217. Copperheads have cross bands (not blotches) that are narrowest middorsally. The timber rattlesnake and massasauga both have rattles on the tail tip. Hog-nosed snakes have pointed, upturned noses. Milksnakes and prairie kingsnakes have smooth scales and undivided anal plates. Juvenile racers have smooth scales and lack the temporal bar.

Additional subspecies

71. The Eastern Fox Snake, *Elaphe vulpina gloydi*, may be found along the eastern and southern shorelines of Lake Huron and the northern and western shores of Lake Erie, where it is associated with the marshes and dunes, and occasionally open woodlands. This snake also utilizes meadows, pastures, and farmlands as habitats. It swims well.

The dorsal and lateral ground color is yellowish, tan, or buff. The back may be somewhat darker than the sides. The dorsal blotches are deep brown to black

71. Eastern Fox Snake

and average about 34 (28–43) in number. The first (nape) blotch is elongate and may be shaped like an H or a U with the open end anteriormost. Smaller lateral blotches alternate with the dorsal blotches. The first blotch on each side of the neck is elongate. The cream to yellow belly is checkered with dark, squared spots. These often extend outward to include the first or first and second rows of lateral scales. The crown and cheeks of the eastern fox snake are often a coppery red. Juvenile and young adults may have a temporal stripe extending from the eye to the back of the mouth, a dark bar across the top of the snout, one along the trailing edge of the prefrontals, and another dark line from below each eye to the lip. These fade with age and the head of older snakes may be pretty much devoid of all markings.

Hatchling eastern fox snakes are paler examples of the adults and lack the coppery head tones.

Adults prey largely on meadow voles and other wild mice.

Kingsnakes and Milksnakes, genus *Lampropeltis*

Four species of kingsnakes and milksnakes occur in eastern and central United States. Three of these have two or more subspecies. Intergrades and color variants also occur.

These snakes are closely allied to the rat and pine snakes. Captives of all three genera have hybridized and produced viable babies.

The eastern representatives of the common kingsnake have 21–25 scale rows, the scales of the milksnakes are arranged in 19–23 rows, and those of the prairie kings are in 21–27 rows. The gray-banded kingsnake has scales in 25 rows.

The scales of all members of this genus are smooth and appear shiny and polished. All species and subspecies have round pupils and undivided anal plates.

The snakes of this genus are noted for their occasional ophiophagous (even cannibalistic) tendencies. Kingsnakes and milksnakes seem immune, or at least very resistant, to the venoms of the venomous snakes on which they may prey. Other prey items include frogs, toads, salamanders, reptile eggs (especially those of turtles), baby turtles, lizards, other snakes (including venomous species), small mammals, and ground-dwelling birds. Neonate kingsnakes also eat some invertebrates.

The prairie and the common kingsnakes have dark ground colors, whereas the milksnakes can be clad in very brilliant colors. The gray-banded kingsnake is intermediate.

Some of the milksnakes are remarkable mimics of the venomous coral snakes but have the ring sequence arranged differently. The harmless milksnakes have the two caution colors (red and yellow) separated by black.

The prairie and common kingsnake groups may be seen by day in areas of open lands, including pastures and agricultural areas, along canals and marshlands, behind loosened bark on dead trees, beneath debris in rural yards, and in woodland clearings. The milksnakes and gray-banded kingsnake are very secretive and strongly nocturnal. The former are often associated with woodlands or prairie lands, the latter with sparsely vegetated escarpments and boulder-strewn aridland habitats.

All members of this genus are oviparous. Breeding and egg laying take place following the snake's emergence from hibernation (if applicable). Males of the gray-banded kingsnake indulge in stylized courtship wrestling bouts, with the winner often breeding the female. The wrestling bouts seem less apt to occur, or are at least less overt, with other kingsnake species. Captives of most kingsnake forms routinely double clutch (have two clutches of eggs annually at intervals of 20–30 days). Apparently, this seldom happens in the wild. Clutches of captive snakes are also often larger than those of wild examples.

Clutch sizes vary: gray-bands, 3–14; prairie, 3–21; eastern races, 4–24; milksnakes, 6–24 eggs. Incubation takes 54–70 days.

Gray-banded kingsnakes and milksnakes are most apt to be surface-active following warm spring and summer thunderstorms, before moonrise or after moonset, or on nights of the new moon. Overcast nights, no matter the moon phase, also induce surface activity. Prairie and eastern kingsnakes are more diurnal but are also stimulated to activity preceding or following storms or during other periods of low barometric pressure.

Kingsnakes and milksnakes are powerful constrictors. Most of them kill their prey before eating it. The gray-banded kingsnake is somewhat less diligent in its constricting practices; if the prey is small enough, it may be swallowed without constriction.

Kingsnakes often readily assume a striking S shape but many are more reluctant to actually bite. However, if the snake is picked up and the head not restrained, it may deliberately choose an area of soft flesh (such as between fingers) to bite, chew, and hold on to.

These snakes often vibrate their tail; in dry vegetation this produces an easily audible and startling whirring reminiscent of a rattlesnake.

As a group, these snakes are very well adapted to their chosen habitats. Their contrasting coloration and patterns, which render kingsnakes so visible when in the open, afford a wonderful camouflage when a few recumbent vines or plant stems are added to the picture. Remarkable though it may seem, when a milksnake is in motion the bright saddles and rings actually tend to obscure the outline of the snake.

72. Gray-banded Kingsnake

Lampropeltis alterna

Disposition: This nonvenomous snake seldom bites even when first caught in the wild.
Abundance/Range: The gray-banded kingsnake was once thought to be a very uncommon species. It has now been learned that, rather than rare, this snake is simply essen-

72a. Gray-banded Kingsnake, bright Blair's morph

72b. Gray-banded Kingsnake, dark Blair's morph

72c. Gray-banded Kingsnake, alterna morph

72d. Gray-banded Kingsnake, alterna morph

tially nocturnal and found in habitats seldom visited by humans. It occurs in southwestern Texas, adjacent New Mexico, and northern Mexico.

Habitat: This is a snake of aridland canyons, boulder fields, fissured escarpments, and creviced road cuts. It is occasionally found crossing desert highways.

Size: The record length for this rather slender serpent is 57¾ inches. The more commonly seen adult size is 24–42 inches. Hatchlings are about 10 inches long.

Identifying features: The gray-banded kingsnake is variable in both color and pattern. There is in fact no question that it is the most variable of our lampropeltines. There are two predominant color morphs, but an immense array of intermediate phases bridge the gap between these. The greatest variability of pattern and color seems to occur about midway in this snake's east to west range. To the west, alterna-phase animals seem to predominate; to the east, the bias is toward the Blair's phase.

Blair's-phase gray-banded kingsnakes were once thought to be a different snake species than the alterna phase. The ground color of the Blair's phase varies from light to very dark gray. It has black-bordered red or orange bands about one-half the width of the gray bands. The red bands are bordered by black, and the black may, in turn, be bordered on its outer edges by a very narrow band of grayish white. The darkest specimens may have the red so suffused with melanin that it is all but invisible or, more rarely, may be lacking entirely.

The alterna phase is a less colorful but no less interestingly patterned snake. This morph usually also has a ground color of variable gray, but may rarely be olive tan or olive brown. The snake's name is derived from its alternating bands of darker color, one a series of spots, the next an unbroken band. The unbroken bands are proportionately narrow and may contain a varying number of orange scales.

Gray-banded kingsnakes that virtually lack a pattern, as well as an occasional striped individual, have also been found in the wild.

Similar species: The Texas lyre snake, another slender, big-headed serpent, looks at least vaguely like some alterna-phase gray-banded kingsnakes. However, the lyre snake lacks orange in its pattern and has vertically elliptical pupils. Some examples of the rock rattlesnake also resemble alterna-phase gray-bands. The rattlesnakes are proportionately stout and lack orange in the pattern. They have elliptical pupils, facial pits, and a tailtip rattle.

73. Prairie Kingsnake

Lampropeltis calligaster calligaster

Disposition: This snake is occasionally quite defensive and may bite.

Abundance/Range: Although seldom encountered even where it is well established, the prairie kingsnake is by no means uncommon. It ranges westward from western Kentucky and northwestern Indiana to southeastern Nebraska, western Oklahoma, and southern Texas.

Habitat: This is a snake of yielding, sandy soils, a substrate in which it may burrow easily. Open mixed and pine woodlands, pastures, meadows, and scrublands are among the habitats utilized by this snake. It is often discovered beneath roadside trash.

Size: This is a kingsnake of moderate length and girth. It is adult at 36–48 inches in length, with a record size of 56 inches. Hatchlings are about 10½ inches long.

Identifying features: The ground color of this variable kingsnake ranges from gray to olive gray, tan, or reddish brown. The 55–65 dark-edged dorsal saddles may be olive brown to tan to reddish, but most often have at least a vague reddish tinge. The blotches may be broadest anteriorly and narrowest posteriorly, or may be about the same breadth from front to back. The blotches may be a single row of saddles or some or all may be broken vertebrally. Vestiges of striping may be visible, especially on old examples that are preparing to shed. A row of smaller blotches of irregular outline is present along each side. Contrast

73. Prairie Kingsnake

prairie
south Florida mole
mole

between blotch color and ground colors may be great or minimal. The belly is paler than the dorsal ground color (often off-white, pale olive, or yellowish) and is usually spotted with squares or rectangles of dark pigment. There is a single dark postorbital stripe. The smooth scales are in either 25 or 27 rows and the anal plate is not divided.

Similar species: The Great Plains and southwestern rat snakes and the corn snake all have slightly keeled dorsal scales and (usually) very prominent checkers on their bellies. Glossy snakes have smooth scales but have unmarked bellies and only 2 (rather than 4) prefrontal scales. The eastern milksnake has a light V or Y on the top of its head.

Additional subspecies

74. The South Florida Mole Kingsnake, *Lampropeltis calligaster occipitolineata*, is the smallest and southernmost race. Most are 18–30 inches in length, rarely attaining a length of more than 40 inches. It is a busily patterned, slender grayish snake with olive overtones. The 75 or more dark olive gray to dark olive green dorsal blotches are usually dark edged. The individual dark scales may have lighter centers. Smaller side blotches alternate between the primary dorsal ones. A postorbital bar is present, but may not be well defined. The sutures of

74. South Florida Mole Kingsnake

the posterior upper labials are dark. Hatchlings are silver gray laterally, olive tan dorsally, and have black dorsal and lateral spots. A series of dark lines *may* be present on the back of the head. The venter is pale, and may be spotted or clouded (but is seldom strongly checkered) with dark pigment. The smooth scales are in 21 or 23 rows and the anal plate is not divided.

Very little is known about the habits of this slender, fossorial snake. Most of the few specimens found have been taken from sandy roads in orange groves or while crossing paved roads on spring afternoons. It is found in the central Florida counties of Brevard, Desoto, Glades, Hendry, and Okeechobee.

Captive females of this mole kingsnake have laid 3–8 eggs. Nothing is known about its reproductive biology in the wild. Hatchlings are about 7 inches in length.

75. The Mole Kingsnake, *Lampropeltis calligaster rhombomaculata*, is adult at 25–36 inches in length, but occasionally attains a length of nearly 48 inches. The ground color may vary from gray through tan to reddish. The dorsal saddles (which can number up to 71) may be olive brown to tan to reddish, but most often have at least a vague reddish tinge. The belly is paler than the dorsal ground color and may be spotted or clouded (but is seldom strongly checkered) with dark pigment. There is a single dark postorbital stripe. The smooth scales are in either 21 or 23 rows and the anal plate is not divided.

Open woodlands (both of mixed tree species and of pine), pastures, meadows, and scrublands are among the habitats utilized by this snake. To date it has been found in two disjunct ranges in the Florida panhandle counties of Madison, Gulf, Calhoun, and Bay, westward to eastern Louisiana, and northward to

75a. Mole Kingsnake (red morph)

75b. Mole Kingsnake (olive morph)

eastern Tennessee and Maryland. It is possible that much of the perceived hiatus in the range of the mole kingsnake is due simply to its secretive nature.

76. Eastern Kingsnake

Lampropeltis getula getula

Disposition: Although it is nonvenomous, the eastern kingsnake can be formidably defensive. Many of these snakes will assume a striking S shape at the first signs of molestation and bite if able. Others bide their time, allowing themselves to be held, pushing with their nose against a finger or arm, then very deliberately biting and chewing.

Abundance/Range: Throughout much of its range, the eastern kingsnake remains a common snake. However, it has completely disappeared from other areas where it was once common for reasons yet unknown. It is active during the day when temperatures are moderate. When the hottest days of summer arrive, the snake spends more time below ground and is less frequently seen. This remarkably powerful constrictor ranges northward from north Florida

76. Eastern Kingsnake

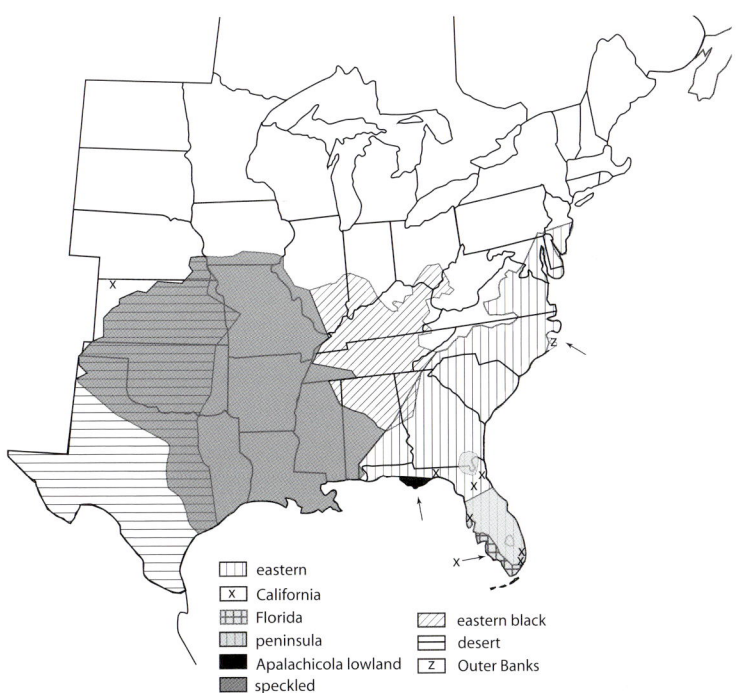

eastern
X California
Florida
peninsula
Apalachicola lowland
speckled

eastern black
desert
Z Outer Banks

throughout most of the Atlantic seaboard states to the Pine Barrens of central New Jersey.

Habitat: Although it is secretive, the eastern kingsnake is drawn by the proliferation of rodents or frogs associated with humans. Look for it beneath human-generated surface debris (boards and roofing tin are particularly favored), near old houses, trash piles, dumps, and barns, along stonewalls, and in other such areas. Eastern kingsnakes are also found along the edges of swamps, marshes, dikes, meadows, fields, and weedy ponds. Sunny railroad embankments and road shoulders, stump holes, and sunny clearings are also utilized.

Size: Although normally 48–60 inches long, some eastern kingsnakes may attain a length of 84 inches. Hatchlings are about 8–11 inches long.

Identifying features: The ground color of this large snake is black or very deep brown. The chainlike dorsal pattern is white, off-white, or cream. There are usually 30–40 dorsal saddles. The chain markings are usually narrow (but in the vicinity of Charleston, South Carolina, can be several scales wide). The smooth dorsal scales are in 17–25 (usually 21) rows and the anal plate is undivided. The light markings of hatchlings may be washed with a variable suffusion of rose or orange. This usually disappears within a few shedding cycles.

Similar species: With its shiny scales, rounded nose, and chain pattern, the eastern kingsnake is not easily confused with any snake except other races of this species. Use range as a diagnostic aid. Also see accounts 80, 82, and 84.

Additional subspecies, intergrades, and color patterns

77. The California Kingsnake, *Lampropeltis getula californiae*, is a popular pet trade subspecies. Although this snake is not yet known to breed in the wild in the eastern or central United States, escaped or released California kingsnakes are now found with some regularity in many areas where they are not native. Although 36–42 inches is the normal length of this snake, some California kingsnakes may near 60 inches. Hatchlings are about 8–11 inches in length.

The ground color of the normal phase of the California kingsnake is black

77a. California Kingsnake, banded morph

77b. California Kingsnake, striped morph

or very deep brown and the snake has a pattern of broad white cross bands. The nose and face are largely white. A second phase lacks cross bands but has a stark white vertebral stripe. The stripe may be entire or broken one or more times along its length. Again, the face is largely white. The smooth dorsal scales are in 23 or 25 rows. Albino and other color/pattern aberrancies are well documented.

78. The Florida Kingsnake, *Lampropeltis getula floridana*, may occasionally attain a length of 66 inches although most adults are 42–54 inches long. It is a variably colored race with a busy, but often muted, pattern. The cross bands number 40 or more. Each dorsal and lateral scale has a yellow(ish) base and a brown(ish) tip. Age-related (ontogenetic) lightening of color can be pronounced, especially in specimens from the light limestone substrates of extreme southern Florida. These lightest specimens were once called Brooks or South Florida kingsnakes, *L. g. brooksi*, and, confusingly, are still referred to in this manner by herpetoculturists.

The sides may, or may not, be lighter than the back. Males from southern Dade and Monroe counties and western Collier and Lee counties are often considerably lighter in color than adult females from the same area. The smooth scales are usually in 23 rows. Hatchlings have a well-defined pattern and are often suffused with orange on their sides.

Look for this large but secretive snake beneath human-generated surface debris (boards and roofing tin are particularly favored), along grassy, frog-laden canal banks, and near old houses, trash piles, dumps, barns, and other places

78. Florida Kingsnake

with adequate cover and food. In defining the range of this kingsnake, we fol-low Blaney (1977). In its pure form, the Florida kingsnake ranges south along Florida's Gulf Coast from Tampa Bay to the tip of the peninsula.

Some populations of the Florida kingsnake prey preferentially on turtle eggs and baby turtles.

79. The Peninsula Kingsnake, *Lampropeltis getula floridana* x *L. g. getula,* is the kingsnake encountered on most of Florida's southern peninsula. Typically 42–54 inches long, occasional specimens measure more than 72 inches.

This is the commonly seen pet-store kingsnake with the milk chocolate to chocolate spots on a cream to butter yellow ground color and a busy pattern. Juveniles are darker than the adults. Peninsula intergrade kingsnakes often re-tain much of the dark pigmentation and the relatively strong pattern of the hatchlings throughout their lives. As would be expected, those at the southern and western periphery of the range are often more like the Florida kingsnake in appearance than those from more interior areas of the range. The smooth scales are in 23 rows.

This kingsnake occurs along canals in the cattle ranches, cane fields, and sod fields of central Florida. The snakes (and their prey) are concentrated in the mazes of oolitic limestone that line the miles of drainage and irrigation canals in that region. The snakes thermoregulate on the ground surface on sunny days in cold or cool weather, but seem to become largely fossorial during the hottest days of summer. Look for them in sandy open areas, in clearings, in the prox-imity of human habitations, beneath debris in trash heaps, and along drainage systems in meadows, pastures, and agricultural areas. As currently understood,

79. Peninsula Kingsnake

the range of this intergrade form extends northward east of Lake Okeechobee from north of Miami to Citrus and Volusia counties. A pocket of what are apparently these kingsnakes occurs in the vicinity of the Osceola National Forest in northeastern Florida and immediately adjacent Georgia.

80. The Apalachicola Lowland Kingsnake, *Lampropeltis getula getula* x *L. getula* subspecies, is a wonderfully variable snake of uncertain parentage. It is typically 42–54 inches in length, but occasional specimens may attain 60 inches. The majority of the kingsnakes from Florida's Apalachicola Lowlands are very different in appearance from the eastern kingsnakes to the west, north, and east of the valley. Over the years the Apalachicola Lowland kingsnakes have been referred to as both "Goin's kingsnake" and "blotched kingsnake." They were once afforded the now invalid subspecific status of *L. g. goini*.

The kingsnakes from the Apalachicola Lowlands are of variable appearance. Typical patterns include (1) a light spot on each lateral and dorsal scale, no cross bands, and a dark vertebral stripe; (2) as mentioned above but with broken horizontal bars on the flanks; (3) as mentioned above but lacking the flank markings and the vertebral stripe; (4) light-spotted dark saddles (of variable intensity and definition) with white bars at least 4, and in some cases 8, scales wide. The result of the wider bands is fewer (20–25) dorsal saddles.

Hatchlings may be suffused with strawberry or orange, strongest laterally. This attractive color usually disappears within several shed cycles.

80a. Apalachicola Lowland Kingsnake, blotched morph

80b. Apalachicola Lowland Kingsnake, striped morph

Studies have attempted to define the genetics of the kingsnakes in this population but have seemingly posed at least as many questions as they have answered. It does seem likely that this snake is an intergrade form, with the eastern kingsnake one known parent species but the other remaining a question. With that in mind we suggest that, for the moment, the snake in question be called the Apalachicola Lowlands color variant.

The smooth scales may be in 21–25 (usually 23) rows. This snake is primarily diurnal but during hot weather may also be active after nightfall. Much of the area in which it is found is so heavily overgrown with cat briar and brambles that the snake would not be visible unless on the roadway. Much about its lifestyle and activity patterns remains speculative. This is a snake of sandy open pine and mixed hardwood areas that support tangled and pervasive undergrowth.

The habitat is typified by the presence of tiny ponds, swamplands, and drainage canals. These kingsnakes are occasionally seen in areas of tall grasses, but more commonly where undergrowth is denser. This color variant occurs in the Apalachicola Lowlands, between the Apalachicola and the Ochlockonee Rivers, and is quite probably the rarest (or at least the most uncommonly seen) of the large kingsnakes.

81. The Speckled Kingsnake, *Lampropeltis getula holbrooki*, is rather well known for its irascible disposition. Occasional specimens are nearly a full 72 inches in length but most are much smaller. Black predominates, but each dorsal and lateral scale usually bears a yellow dot of variable intensity. The chain pattern so typical of the eastern kingsnakes is usually at least weakly visible. This subspecies intergrades with other races where ranges abut. The venter is usually predominantly black, with variable spots and bars of yellow.

Hatchlings are often somewhat more contrastingly patterned than the adults.

The speckled kingsnake has a proportionately narrow head and usually 21 rows of smooth, shiny scales.

Although they are essentially terrestrial, speckled kingsnakes can and do swim well and occasionally climb. Within their habitat they utilize all manner of ground cover, and are often found, sometimes in numbers, beneath boards, roofing tin, and other such debris. During cool weather kingsnakes are primarily diurnal, sometimes sunning in rather exposed areas but more often secluding themselves beneath recumbent vines and grasses, where their color and pattern melds magnificently with the traceries of sun and shadows. As the weather warms, these snakes become more prone to crepuscular activity. This race ranges from southeastern Alabama to eastern Texas and northward to southern Iowa.

81. Speckled Kingsnake

82. Eastern Black Kingsnake

82. The Eastern Black Kingsnake, *Lampropeltis getula nigra*, is the darkest of the eastern races. It is adult at 36–48 inches in length; occasional specimens approach a full 60 inches. It inhabits open woodlands, woodland clearings and edges, meadows and their edges, ditch, canal, and pond edges, and the edges of swamps, marshes, and impoundments. The edges of dumps and trash piles are especially favored sites as are the sunny, grassy slopes of embankments and dikes. Some moisture, ample hiding areas, and a rather constant food supply seem to be the only criteria needed by kingsnakes. It ranges from southeastern Illinois eastward to western West Virginia and southward to central eastern Georgia.

Black predominates, especially dorsally. Most specimens have a variable amount of yellow on the sides. In many cases the yellow chain pattern so typical of the eastern kingsnakes is at least weakly visible. On some examples, though, especially old individuals, the pattern is all but invisible, being present only as a few yellow spotted scales that ascend the sides. The venter is usually predominantly black, with variable spots and bars of yellow.

Hatchlings are often somewhat more contrastingly patterned than the adults.

83. The Desert Kingsnake, *Lampropeltis getula splendida*, attains 60 inches in length but usually no more than 42–51 inches. It ranges widely in arid areas from central Texas to southeastern Arizona and northward only to central New

83. Desert Kingsnake

Mexico. It also occurs in Mexico. It intergrades readily with the speckled king-snake along its eastern range periphery. This interesting snake has black and yellow speckled sides and chainlike yellow dorsal bars separating large black (or dark brown) dorsal saddles. The venter is predominantly black, with variable spots and bars of yellow.

Hatchlings are often somewhat more contrastingly patterned than the adults.

This kingsnake has 23 or 25 rows of smooth, shiny scales.

Look for the desert kingsnake in riparian situations, near stock tanks, or along irrigation canals. It also wanders far into desert situations. This species is usually nocturnal in its pattern of surface activity.

84. The Outer Banks Kingsnake, *Lampropeltis getula sticticeps,* is of question-able taxonomic validity. Confined in distribution to the Outer Banks of North Carolina and the immediately adjacent mainland, the selection of habitat by this kingsnake morph seems identical to that chosen by the mainland form. Dumps, trash piles, embankments, dikes, and water edge situations are all utilized. The Outer Banks kingsnake is listed as a species of special concern by the state of North Carolina.

The designation of this snake as a subspecies has been fraught with uncer-tainty since its description in 1942. Over the years, its recognition was based on an enlarged, tapering, rostral scale, a ground color lighter than that of the mainland snakes, fewer than 35 white bars, much white on the head and in the dark fields on the sides, and ventral scales numbering 210 or fewer. These

84. Outer Banks Kingsnake

characteristics occasionally appear on eastern kingsnakes, yet some researchers continue to consider this a valid subspecies.

85. Eastern Milksnake

Lampropeltis triangulum triangulum

Disposition: The eastern milksnake will bite if handled carelessly or otherwise provoked.

Abundance/Range: Seemingly rare in some areas and abundant in others, the perceived scarcity of the eastern milksnake in some regions may stem from the snake's inordinate ability to hide in even minimal cover. For example, about a dozen adults were found one spring morning beneath a single weathered 4 x 8-foot sheet of plywood in rural northwestern Massachusetts. Despite extensive searching, milksnakes had not been previously noted in the immediate area, nor were they found again. This snake ranges southward from Michigan's Upper Peninsula, southern Quebec, and southeastern Maine to the western Carolinas. It intergrades with both the red milksnake and the scarlet kingsnake in some areas where ranges abut.

Habitat: The eastern milksnake occurs in a wide variety of habitats, from mountaintop to coastal plains. Like most members of the genus, eastern milksnakes are secretive; although they can be quite numerous, they may go mostly unnoticed.

Milksnakes may be encountered beneath manmade debris and litter at the edges of dumps or trash piles. They also are found on rocky hillsides where they seek seclusion under rocks, in field and meadow edges where mats of vegetation serve as shelters, and on road shoulders, damp lowlands, and other such

85. Eastern Milksnake

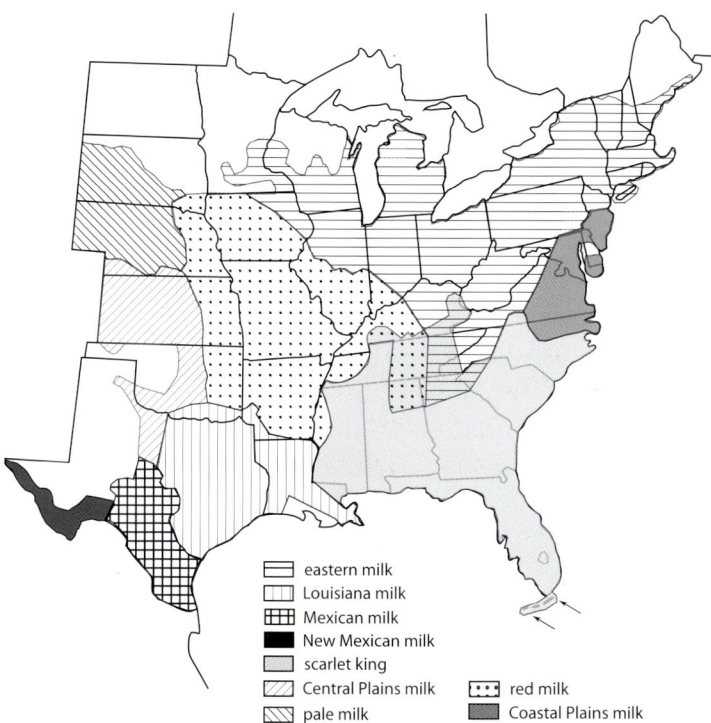

eastern milk
Louisiana milk
Mexican milk
New Mexican milk
scarlet king
Central Plains milk
pale milk
red milk
Coastal Plains milk

areas where their amphibian, reptile, and small rodent prey is abundant. These snakes are occasionally seen around buildings in barnyards or near deserted and tumbling down dwellings where mice are numerous.

Size: Most adults of the eastern milksnake measure 24–42 inches in length. The record verified size was 51 inches. This is a slender snake with a moderately distinct head. Hatchlings measure about 8½–10½ inches long.

Identifying features: Although most babies are actually brightly colored, these snakes often dull noticeably with advancing age.

Hatchlings have dark-edged red saddles on a pale gray ground color. Red lateral spots alternate between the dorsal saddles. This pretty and highly contrasting combination is retained by most juveniles and by some adults. Usually, though, with growth, the red saddles dull to maroon or olive red, or may, in some cases, actually take on a decidedly olive, cinnamon, or brownish sheen. The ground color darkens to dark or olive gray. The top of the head has much red, the tip of the snout is gray. A dark line extends posteriorly from the eye to the back of the mouth. The first dark nape spot is variable, but often extends forward onto the top of the head and contains in its center a light area in the shape of a V or a Y. However, this spot may occasionally stop abruptly at the back of the head, allowing the presence of a light collar. The belly is variably checkered (sometimes blotched) with dark bluish black on a gray ground color. The dark ventral spots may extend upwards onto the first row of body scales.

The smooth scales are often in 21 rows, but may vary from 19 to 23. The anal plate is undivided.

Similar snakes: Corn snakes have weakly keeled scales dorsally and a divided anal plate. Copperheads (for which this species is often mistaken) have unmarked, coppery colored heads and cross bands that are narrowest middorsally. The prairie kingsnake can be confusingly similar, but its posterior lateral spots are often irregular or poorly defined. The dorsal saddles of the mole kingsnake are usually more widely separated and the lateral spots may be weak or lacking. Coral snakes are ringed and have the two caution colors, red and yellow, touching.

Additional subspecies and color variants

86. It is the Louisiana Milksnake, *Lampropeltis triangulum amaura*, that is found from eastern Louisiana to central Texas and northward to southern Oklahoma.

86. Louisiana Milksnake

This subspecies inhabits wooded areas where it utilizes rocks, ground debris, and the bark of dead trees as cover. This small snake seldom exceeds 24 inches in length. The record size is a slender 31 inches. The black and white rings cross the belly, but the red may not, being separated midventrally by black pigment. The snout is usually gray or white.

87. With a record size of a stocky 41½ inches, the Mexican Milksnake, *Lampropeltis triangulum annulata*, is one of the larger and more beautiful of the ringed snakes. It ranges southward from central Texas into Mexico. Both the red and yellow (or white) colors may be separated midventrally by black pigment. The black rings are usually widest vertebrally but do not narrow significantly on the sides. The yellow bands (rings) are usually of similar diameter throughout. The 14–26 red bands (rings) are quite wide. The snout is mostly black but there may be some white on the anterior upper labial scales. This is a species of sandy, open, semi-arid areas, and is often encountered near a water source or while crossing roadways in warm weather.

87. Mexican Milksnake

88. The New Mexican Milksnake, *Lampropeltis triangulum celaenops*, ranges westward from Trans-Pecos region of Texas to central New Mexico, then northward to southern Colorado. Disjunct populations that most closely fit this race's description are now being found in central western Arizona. The black snout of this pretty milksnake may or may not be flecked with white. The 17–30 red bands are narrowest middorsally and are interrupted midventrally by black pigment. The belly is predominantly yellow (or white). This is a slender milksnake that barely attains a maximum size of 25 inches but is rarely longer than 20 inches. It may be found well up into the mountains as well as on sparsely

88. New Mexican Milksnake

wooded, often arid, lowlands. It is often found near a water source or at night as it crosses a road.

89. The Scarlet Kingsnake, *Lampropeltis triangulum elapsoides,* is the most divergent of the milksnakes in both appearance and habits. This is the tricolored snake that would be most readily confused with the venomous coral snake. However, the two are easily differentiated: the two traffic signal warning colors—red and yellow, which touch on the North American coral snakes—are separated by black on the scarlet kingsnake. This snake is adult at 14–18 inches in overall length. While the largest specimens are in the 27–29-inch range, any measuring 20–24 inches are considered huge. Hatchlings are a slender 5¾ inches long.

Scarlet kingsnakes are common to abundant in areas with suitable cover.

89. Scarlet Kingsnake

They are often associated with rather well-drained pinelands, but occur in mixed woodlands as well. In the spring these brightly colored snakes often secrete themselves behind the loosened bark of dead but still standing and fallen pines. They also occur in rotting stumps and beneath trash. The scarlet kingsnake occurs throughout Florida, then ranges northward and westward to North Carolina, Tennessee, and extreme eastern Louisiana.

The smooth scales of the scarlet kingsnake are arranged in 19 rows. The brilliant bands of red, black, and yellow (or white) completely encircle the body. The red and the yellow bands are separated by black bands. The nose is usually red. Pallid (pastel) individuals are occasionally encountered. Except for having the yellow replaced by white, scarlet kingsnake hatchlings are colored like the adults. The head is proportionately very narrow, the snout is narrowly rounded. The anal plate is undivided.

90. The Central Plains Milksnake, *Lampropeltis triangulum gentilis*, is common to abundant from eastern Kansas and southern Nebraska to eastern Colorado and northern Texas. It can be abundant on rocky hillsides, canyon sides, and along tree lines. Like other milksnakes, it is often found near a water source. This snake is variable in color, but the red is usually paler than on the eastern races. The red and the white (rarely yellowish) markings may be about the same width or the red may be noticeably wider. There are 20–39 red bands that are narrowly separated midventrally by black. The white coloration is in the form of complete rings. The black is the narrowest. The snout is usually gray or white. Although usually smaller, this milksnake occasionally tops out at a length of 36 inches.

90. Central Plains Milksnake

91. Pale Milksnake

91. The Pale Milksnake, *Lampropeltis triangulum multistriata*, ranges westward from eastern South Dakota to central Montana and central Wyoming, where it is often associated with rocky hillsides, open woodlands, and streambeds. Although adults of this race measure 16–26 inches, occasional examples can attain 30 inches or slightly larger. Hatchlings measure about 9 inches in length. While a beautiful snake, the pale milksnake is not quite as brilliantly colored as more easterly races. The light bars are often gray(ish) rather than white or yellow, the black is reduced in quantity, and the red bands are muted to orange red or dusty red. The top of the head is dark and the snout is orangish with scattered dark scales. The bands do not completely encircle the body, leaving the midventral area (usually) unpatterned light gray to white.

The body scales are smooth and usually in 19 rows.

92. The Red Milksnake, *Lampropeltis triangulum syspila*, attains a length of 20–30 inches, with a record at 42 inches. Hatchlings are about 9¾ inches in length. This is a milksnake of open woodlands, rocky hillsides and meadows, fallow fields, pastures, and farmlands. It may be quite common where surface cover such as tumbling buildings and trash piles exist. The range of the red milksnake encompasses the heartland of the United States. It is found throughout Missouri, most of Iowa, and Arkansas, as well as in southeastern North Dakota, eastern Kansas, northeastern Oklahoma, southern Illinois, southwestern Indiana, the westernmost regions of Kentucky and Tennessee, and northwestern Louisiana.

92. Red Milksnakes

This is a saddled rather than a ringed milksnake. The ground color is off-white to pale gray and the black-edged markings are a bold and bright red that extends to, or nearly to, the ventral plates. Lateral spots are absent or small. The top of the head is red; the snout is gray. The belly coloration is of black and white checks.

The scales are smooth, in 21 or 23 rows.

93. The so-called Coastal Plains Milksnake is a problematic form that most taxonomists now consider an intergrade between the eastern milksnake and the scarlet kingsnake. It was once designated *L. t. temporalis*. This snake rang-

93. Coastal Plains Milksnake

es from the Pine Barrens of New Jersey southward to northeastern North Carolina.

The Coastal Plains milksnake has saddles of red outlined with black rather than rings of color. The dorsal saddles extend far down the sides, often reaching, or at least nearing, the ventral plates. There are no lateral blotches. Rather than a variably elongate blotch, the nape marking is in the form of a precise collar. The top of the head is red, whereas the snout anterior to the eyes and the sides of head are light. The light bands can vary from light gray to pale yellow. The entire belly may be variably checkered, black on white, may be unmarked midventrally, or may be of intermediate pattern. This snake attains 20–40 inches in length. The smooth scales are in 21 or 23 rows.

Gopher Snakes, Pine Snakes, and Bullsnakes, genus *Pituophis*

The members of this genus are moderately large to large constricting snakes with heavily keeled scales. All have a rather belligerent nature. Thanks to a glottal modification these snakes have the ability to hiss loudly. They also have a penchant for vibrating their tail. Many pine and gopher snakes will strike and bite savagely if closely approached. The members of this genus have proportionately narrow and elongate heads. The rostral scale—the scale on the tip of the snout—is protruding, strongly convex, and much higher than wide. These snakes have 4 prefrontal scales. The body scales are in 27–33 rows at midbody and the anal plate is undivided.

These snakes are accomplished burrowers (gopher snakes seem less inclined to burrow than pine snakes) and spend much of their time below ground. They pursue their rodent prey (often pocket gophers) in their underground burrows.

Underground chambers (often of the snake's own making) are used for egg laying. Most clutches consist of 4–12 (occasionally to 27) comparatively large eggs. Communal nesting is well documented in some species. Incubation takes 58–90 days. As might be surmised, the hatchlings are also quite large, some exceeding 20 inches in length.

These snakes can climb, but do not seem to do so often. During warm weather bullsnakes and gopher snakes are commonly seen on ranchlands and crossing desert roadways after dark. Pine snakes are less frequently seen, generally roaming most widely during the spring breeding season.

These secretive snakes are associated with sandy, yielding soils.

It is now rather generally conceded that there are either two or three species of this genus in the United States.

A variety of small mammals (gophers, moles, rabbits, rats, mice) as well as birds and their eggs are eaten. Hatchlings are also known to eat lizards. When gophers or rats are encountered in the rodents' underground warrens, pine snakes apparently forge on past the rodents and pin them against the wall of their tunnel with a muscular arc of the snake's body.

94. Sonoran Gopher Snake

Pituophis catenifer affinis

Disposition: This, like other members of the genus, will become actively defensive if frightened. It will bite, hard and repeatedly.

94. Sonoran Gopher Snake

Sonoran gopher
Louisiana pine
X bull

Abundance/Range: This is a commonly encountered snake. It seems to be less fossorial than the related pine snakes. It ranges westward from West Texas to southeastern California, and southward from southwestern Colorado to central Mexico.

Habitat: This is a snake of aridlands. It is found in dry, climatically hostile deserts, dry grasslands, pastures, and rural fields. It is often found on the outskirts of large cities and may be very common in irrigated agricultural areas. It is an able burrower, but also seeks the comparative temperature comfort of rodent burrows.

Size: This snake is among the largest of the North American species. Sonoran gopher snakes 48–72 inches long are not uncommonly found. The record size is 92 inches. Hatchlings vary between 12 and 18 inches in length.

Identifying features: The ground color of the Sonoran gopher snake is tan to yellowish, colors that blend well with the normally sandy background of its habitat. The dorsal saddles and the lateral blotches are darkest anteriorly and posteriorly. These are reddish black on the neck, black on the posterior body and tail, and reddish brown between. A poorly defined brown interorbital bar and a diagonal bar from the rear of the eye to the angle of the mouth are usually visible. The white, off-white, or yellowish venter is variably spotted with dark pigment. Hatchlings and juveniles are very similar to the adult, but have a lighter ground color, hence more contrasting dark markings.

The body scales are heavily keeled and in 31 or 33 rows. The rostral scale is enlarged but not excessively so. The anal plate is undivided.

Similar species: The bullsnake is very similar but has a larger rostral scale and the dorsal blotches are reddish brown all the way to the tail. Hog-nosed snakes are short and stout and have upturned, not bulbous, rostral scales. Rat snakes and kingsnakes lack modified rostral scales.

Additional subspecies

95. The Louisiana Pine Snake, *Pituophis catenifer ruthveni*, occurs in central western Louisiana and adjacent Texas and is apparently very rare. It occurs in rapidly drained, sandy, longleaf pine or pine-oak woodlands with loose soils. It attains an adult size of 42–60 inches, with a record size of 70½ inches. Hatchlings vary between 16 and 22 inches long.

Despite being called a pine snake, this rare serpent is much more like a bullsnake in appearance than any of the other pine snakes. A recent treatise has, without comment, included the Louisiana pine snake as a subspecies of the bullsnake. We have elected to follow suit.

The ground color of the Louisiana pine snake is yellowish. The dorsal saddles and the lateral blotches are dark brown, reddish brown, to almost black. Darker ground color between the anterior blotches, both dorsal and lateral, makes them seem poorly defined. As the ground color lightens posteriorly, the saddles and blotches become much more distinct. The saddles on the rear two-thirds of the body contrast most sharply and cleanly with the ground coloration. The

95. Louisiana Pine Snake

yellowish venter is variably spotted with dark pigment. There are usually small dark spots atop the head, and a variably distinct interorbital bar joining the front of the supraocular scales and continuing posterior to the eye to the angle of the jaw. Hatchlings and juveniles are very similar to the adult, but have a lighter ground color, hence more contrasting dark markings.

The body scales are heavily keeled and in 27–33 rows. The rostral scale is noticeably enlarged, and the anal plate is undivided.

Very little is known about the reproductive biology of this snake either in the wild or in captivity. The clutches seem to be extraordinarily small, containing 1–5 very large eggs. The incubation period is 58–66 days.

96. The Bullsnake, *Pituophis catenifer sayi*, is an abundant inhabitant of a wide range of habitats from southern Alberta and western Wisconsin to northeastern Mexico. This habitat generalist occurs in prairies, deserts, wooded canyons, pasturelands, riparian areas, and ranchlands.

The ground color of the bullsnake is tan to yellowish. The dorsal saddles and lateral blotches are darkest anteriorly but quickly shade to brown, olive brown, or reddish brown and remain that color for most of the length. It is only on the tail tip that they again darken to near black. A poorly defined brown interorbital bar and a diagonal bar from the rear of the eye to the angle of the mouth are usually visible. The white, off-white, or yellowish venter is variably spotted with dark pigment. Hatchlings and juveniles are very similar to the adult, but have a lighter ground color, hence more contrasting dark markings.

96. Bullsnake

The body scales are heavily keeled and in 27–37 rows. The rostral scale is noticeably enlarged. The anal plate is undivided. The record size for this heavy-bodied snake is 102 inches.

97. Northern Pine Snake

Pituophis melanoleucus melanoleucus

Disposition: This is a nonvenomous snake, but it should be noted that the term nonvenomous in no way means docile! It will strike and bite readily.

Abundance/Range: Although the northern pine snake is specifically protected only in New Jersey and Tennessee, it is also protected by Georgia's blanket protection of harmless indigenous snakes in that state. This snake does not seem common at any point in its curiously disjunct range. It occurs in the New Jersey Pine Barrens, in isolated colonies in north central Virginia and central and western Kentucky, in three well-separated ranges in eastern, central, and western Tennessee, in eastern Alabama and northern Georgia, and over all of South Carolina except the southeastern Low Country.

Habitat: This is a species of rapidly drained, sandy uplands and barrens. These snakes are nearly always in the proximity of, or actually in, pine or pine-oak woodlands. The northern pine snake is partial to loose soils in which it may easily burrow. It also uses rodent burrows extensively and spends much of its time below ground.

Size: The northern pine snake is one of the largest of the eastern snakes, often attaining 66 inches in length. The record size is 72 inches. Because of its proportionately heavy build the snake appears even larger than it actually is. Hatchlings are usually more than 18 inches long.

97. Northern Pine Snake

northern
black
Florida

Identifying features: It is in the Pine Barrens of New Jersey that the northern pine snake displays its most highly contrasting patterns. There, the ground color may be nearly white, off-white, pale gray, yellowish, or, rarely, reddish. The dark dorsal saddles (which may contain some light-centered scales) and lateral spots are of irregular outline and dark brown (often this color near the and on the tail), gray, or charcoal in color. The saddles on the rear two-thirds of the body contrast most sharply and cleanly with the ground coloration. The anterior one-third of the body, and especially the neck area, may be dusky because of a suffusion of melanin on the individual scales and interstitially.

In more southern populations, the ground color is more likely to have a yellowish tinge and the dark blotches are of some shade of brown rather than black. The dark blotches contain from one to many light-centered scales. The overall appearance is of a dingier, less contrastingly colored snake.

Examples from South Carolina are dingier still, and usually have a suffusion of dusky pigment anteriorly and even less strongly contrasting blotches, most of which have at least some light-centered scales within.

The head has some dark patterning in the form of dark pigment at the scale edges and sutures or tiny dots on the scales. A dark interorbital line may be present. Dark pigment may be present (especially on southern specimens) in the sutures of the upper labial scales. The venter is unpatterned and white, yellowish, or, rarely, reddish.

Hatchlings have a whiter (rarely off-white or pale pinkish) ground color and more sharply defined blotching.

The head is narrow, the rostral scale is noticeably enlarged, and the anal plate is undivided.

Comments: Heavily gravid females of this impressive snake have an immense girth. Deposition occurs in side chambers off a main burrow constructed by the female snake. The same nesting chamber may be used for many years on end, and communal nesting is well documented. The burrow may be constructed in a sunny spot on a sandy embankment, sometimes beneath boards or logs. Heavily shaded areas are not utilized for underground deposition sites because the ground stays too cool.

Because they are so secretive throughout much of their range, it is not yet known with certainty whether northern pine snakes are merely adept at remaining out of sight or truly rare. Of the large, constricting snakes of the East, the pine snakes are unquestionably the most fossorial. Except in the New Jersey Pine Barrens, where they have been studied extensively, their life history is poorly understood. Some examples of northern pine snakes, surprised in the open, react so adversely to even distant visual stimuli by hissing and rattling their tail that they actually draw attention to themselves when they would be otherwise overlooked.

Additional subspecies

98. The Black Pine Snake, *Pituophis melanoleucus lodingi*, is one of many snakes that is known far better in captivity than in the wild. The habitat of this impressive snake is closely associated with sizable forests of first and second growth longleaf pine. The range of the black pine snake includes southwestern Alabama, southeastern Mississippi, and Washington Parish, Louisiana. The snake's record length is 76 inches. Although the pattern is very similar to that of the northern pine snake, the overall coloration is much darker. The pattern of dorsal blotches and lateral spots is often most visible on hatchling and juvenile snakes. As age and growth progress, a suffusion of melanin all but obscures, or sometimes actually does obliterate the pattern of this black to brownish black snake. The venter is dark, but somewhat lighter in color than the dorsum.

Where the range of the black pine snake and the Florida pine snake abut on

98. Black Pine Snake

Florida's western panhandle county of Escambia, west of the Escambia River a few intergrades between these two magnificent snakes have been found, with one example in captive care now more than 104 inches in length. They are of intermediate color, far darker than a Florida pine snake but light enough to allow at least vestiges of a pattern to be seen.

99. The Florida Pine Snake, *Pituophis melanoleucus mugitus*, has the most diffuse dorsal markings of any of the pine snakes and occasionally attains 90 inches in length. The ground color may be nearly white, off-white, pale gray, or yellowish. The dorsum is usually patterned with gray to brown blotches that are of irregular outline and may contain some light-centered scales. The saddles on the rear two-thirds of the body contrast most sharply and cleanly with the ground coloration. Dark lateral blotching is usually at least faintly indicated posteriorly. Dark markings may be lacking entirely.

The head has some dark patterning in the form of dark pigment at the scale edges and sutures or tiny dots on the scales. A dark interorbital line may be present. Labial scale sutures are usually dark. The venter is unpatterned, white or off-white. Hatchlings usually have a very light ground color and well-defined blotching. This seldom-seen snake ranges throughout all except southern Florida, immediately adjacent Alabama, southern Georgia, or southeastern South Carolina.

These are very secretive snakes that apparently spend most of their time in

99. Florida Pine Snake

burrows of their own making or in the burrows of small mammals. Males (more rarely females) are occasionally found actively crawling on the ground surface in the spring of the year.

Long-nosed Snakes, genus *Rhinocheilus*

This is a genus of beautiful and often busily patterned tricolored snakes. Long-nosed snakes are associated with aridlands and semi-aridlands and are secretive and capable burrowers. They may be found among sheltering rocks, in crevices, in rodent burrows, under debris, or otherwise hidden by day. They prowl widely in the evening and after dark, especially during or following rains.

Long-nosed snakes are kingsnake relatives. If provoked they may either hide their head in their body coils, or assume an S shape and strike animatedly. Juveniles are especially apt to bite if molested. If severely stressed, long-nosed snakes writhe and may autohemorrhage (spontaneously bleed) from the cloaca while voiding musk and feces. Nasal hemorrhaging has also been noted. The precise mechanisms of this behavior are unknown. A frightened long-nosed snake may also vibrate its tail tip.

Long-nosed snakes are efficient constrictors, but often consume small prey items without constricting. While lizards seem the food of choice, small rodents and nestling birds may also be eaten.

A clutch of 4–9 large eggs is produced annually. The incubation period is 60–80 days. The head of this snake is slender and the nose is pointed. The smooth scales are in 23 rows and the anal plate is undivided.

100. Texas Long-nosed Snake

Rhinocheilus lecontei tessellatus

Disposition: This nonvenomous snake may strike and bite if molested or carelessly restrained.

Abundance/Range: This common snake ranges southward from southern Oklahoma and adjacent Colorado through most of New Mexico and the western two-thirds of Texas. From there it ranges far southward into Mexico.

Habitat: The long-nosed snake either burrows or uses the burrows of rodents during the hours of daylight. However, it wanders widely from dusk until long after nightfall. It thrives where lizards are common in dry prairies as well as in rocky arid and semi-arid areas. It may occasionally be found beneath surface debris and rocks.

Size: Most examples found are in the 20–36-inch range. The record size is 41 inches. Large examples are of robust build. Hatchlings are a slender 7–10 inches in length.

Identifying features: The Texas long-nosed snake is a tricolored species with a particularly busy and variable pattern. On adults the more precisely delineated black and red markings are in the form of dorsal saddles. These are separated by white markings a half-scale in width. The sides are strongly speckled with black and white and red and white. The snout may be reddish or whitish, but is often strongly speckled with black. The head is narrow, making the nose appear long. Some adults are very pallid, appearing pink when in motion. The belly is usually white, but dark markings may encroach from each side. Hatchlings and juveniles are a more precisely patterned black and red. The characteristic white speckling develops quickly with growth.

100. Texas Long-nosed Snake

Similar species: The various milksnakes within the range of the Texas long-nosed snakes all have at least some well-developed areas of black on the belly and lack the strongly speckled lateral pattern so typical of the latter. Coral snakes are strongly ringed and have a black nose. Ringed-phase ground snakes are not tricolored and lack the strong speckling.

Short-tailed Snake, genus *Stilosoma*

The short-tailed snake is a small and extremely attenuate kingsnake relative. It is a rarely seen, burrowing Florida endemic, found in well-drained sandy pinelands on the central western peninsula. It is a weak constrictor. The preferred, and perhaps only, food is crowned snakes (*Tantilla*).

These are excitable little snakes that vibrate their tail if startled and will hiss and strike when further annoyed. They are entirely nonvenomous. Very little is known of their life history, other than they are oviparous and produce elongate eggs.

This is one of the most poorly understood snakes in eastern North America. It is usually found when heavy rains saturate the soil and flood its burrows. It is occasionally seen above ground in the late afternoon or evening in midspring. Since this is breeding time for most Florida snakes, and many *Stilosoma* found above ground are males, we can speculate that the prowling snakes are seeking or following the pheromone trails of receptive females.

101. Short-tailed Snake

Stilosoma extenuatum

Disposition: When frightened this slender snake draws its neck into an S and strikes persistently. It also vibrates its tail, producing an audible sound when in dry grasses or leaves.

Size: Most examples seen are in the 15–18-inch range. This snake occasionally attains 24 inches in length.

Identifying features: This very slender (about the diameter of a pencil), small-headed, small-eyed snake is a relatively weak constrictor. It will coil around and partially immobilize its prey, but does not kill it prior to ingestion.

There are two color phases, both with a gray lateral ground color and darker lateral and vertebral spots. One phase also has a gray back, but the other has an orange color between the vertebral spots. The ground color of the dark spotted belly is also gray. The head is barely wider than the neck. The smooth scales are in 19 rows and the anal plate is not divided.

Abundance/Range: The short-tailed snake is a Florida endemic. It is a seldom-seen and probably very rare snake that is fully protected by the state of Florida. In our several decades in Florida, we have found only four specimens and only one of these was actively crawling.

Habitat: The short-tailed snake burrows in the loose substrate of Florida's remaining sandhills. Perhaps its most important area of remaining range is the Ocala National Forest, but the short-tailed snake persists from Tampa on the west coast and the Lake Wales Ridge in Florida's interior to Columbia County.

Similar species: Juvenile racers, coachwhips, and rat snakes have a pattern of

101. Short-tailed Snake

dorsal saddles but are bigheaded and big eyed. Although the pattern of the short-tailed snake is echoed by the two Florida races of mole kingsnakes, the extreme slenderness of the short-tailed snake should provide immediate positive identity.

Water Snakes, Garter Snakes, and allied species, subfamily Natricinae

Eight genera of natricine snakes occur in eastern and central North America. Besides the very typical and well-known water snakes and garter snakes, there are the specialized crayfish snakes, the secretive brown and red-bellied snakes, the burrowing earth snakes, and the tiny water snake look-alikes, the swamp snakes and the Kirtland's snake. A tiny garter snake look-alike, the lined snake, is widespread in the Plains states.

Most of these snakes are associated in some manner with water. Many live either in or over it, or their prey lives in the water. The brown, lined, Kirtland's, red-bellied, and earth snakes are exceptions. They burrow in the earth, sometimes in surprisingly dry locations.

The snakes in this subfamily vary in size from only about 6 inches to a heavy-bodied 6 feet. None are powerful constrictors; indeed, few even attempt to coil around a prey animal when it is seized. The natricines of North America feed upon ectothermic prey such as fish, amphibians, worms, or other invertebrates.

Many natricines have some manner of toxin in their saliva. Bites from some of the North American forms have caused redness and edema, but no serious side effects. Bites from some Asiatic natricines have been blamed for an occasional human fatality.

All North American members of this family give birth to live young.

Once among the most abundant of snakes, the garter and water snakes now seem to be in at least a moderate decline and some have become rare.

Many natricines will bite viciously if carelessly restrained. A few of the darker water snakes may be mistaken for the venomous cottonmouth. Use care when seeking and handling these snakes.

Water snakes at home

The lingering rays of the newly set sun were still painting the outlines of the accumulating puffs of towering cumulus with an unearthly gilt. Green treefrogs chorused in the distance, but the riverbank stridulations of North Florida katydids almost overpowered the calls of the anurans. Twilight didn't linger. Hastened by the buildup of clouds, darkness was soon upon us. A barred owl voiced its characteristic eight hoots and was answered by another owl nearby. Our tiny motor sputtered and died. Used to the irregularity of the motor's compliance with our wishes, Dick and I dug out the paddles and aimed the canoe toward the shore. Mayflies danced ephemeral tangos, their wings gossamerlike in the beams of the flashlight.

Our car was in a pull-off about a mile upriver, so we had just a short way to go. The southern moon was rising but obliterated by a leaden sky. Paddling slowly, we scanned the overhanging limbs and shrubs and were soon rewarded--as the rain began to fall--by the sight of a big midland water snake draped about three feet above the water.

As we continued through the rain we soon encountered a second midland water snake and then a third. By the time we reached the pullout our count stood at about 20, plus a couple of map turtles. We were dripping wet but happy--happy at the finding of the reptiles, but happier yet to finally stow the canoe and gain the dryness of the car.

Kirtland's Snake, genus *Clonophis*

This genus contains but a single species, the little Kirtland's snake of the Great Lakes states (Illinois and immediately adjacent northeastern Missouri, Indiana, Ohio, Michigan, and western Pennsylvania). It was once thought to be a water snake.

From 3 to 10 (rarely to 15) babies are born usually in late summer or very early autumn.

To deter predators (including inquisitive humans) a startled Kirtland's snake may indulge in one or more of a number of ploys. It may flatten its body, writhe energetically, strike and bite; it may become rigid; or it may attempt to flee. As with most natricines when handled, Kirtland's snakes will smear musk, urates, and feces on the hand of their captor.

Thinking that Kirtland's snakes were of rare occurrence, we were astonished when a friend showed us a dozen or so in a tiny vacant lot in inner Cincinnati, Ohio, in 1998. The Kirtland's snakes were intertwined with eastern garter snakes beneath tiny bits of tin and paper. Nevertheless, populations of Kirtland's snake do appear to be declining in many areas of their range.

These snakes feed primarily on earthworms, but slugs may also be accepted.

The keeled scales are in 19 rows; the anal plate is divided.

102. Kirtland's Snake

Clonophis kirtlandi

Disposition: Kirtland's snake can seldom be induced to bite and is entirely harmless to humans.

Abundance/Range: Kirtland's snake was once far more common than it is now. It is such a secretive species that current-day researchers are having difficulty in making meaningful population assessments. For lack of better knowledge, Kirtland's snake is currently considered endangered in its remaining Michigan and Pennsylvania range and threatened in Indiana and Illinois. It is protected in Ohio, but not yet protected in Kentucky or Iowa (if it continues to exist in the latter state).

Habitat: Kirtland's snakes are still found, sometimes in numbers, in urban and suburban areas, further fueling the enigma of the snake's true status. In such urban and suburban situations, *Clonophis* may be found beneath rocks, boards, discarded newspapers, and even cardboard. It is equally secretive in natural

102. Kirtland's Snake

situations, hiding beneath all manner of surface debris and utilizing the burrows of rodents and crayfish.

Size: This snake is adult at 16–18 inches, with the greatest recorded size being
only 24½ inches. Neonates are about 5¾ inches in length.

Identifying features: Except for a single rarely seen variation, the color and pattern of this beautiful little snake is quite standardized. The most commonly seen
dorsal pattern has 4 rows of alternating dark spots, 2 on each side. The ground
color varies from olive tan to dark brown. A tan or brown vertebral stripe may
be visible. The top of the head is black(ish).

Rarely the 2 rows of spots on the upper sides are of reduced size and the
vertebral stripe is wider and prominently defined.

The chin is white, brightening to pale orange under the neck. This intensifies
to a bright reddish orange for the entire length of the belly and tail. A definitive
black dot (by which the species may be easily identified) is present on the outer
edge of each ventral scute. The scales are keeled, in 19 rows, and the anal plate
is divided.

Similar snakes: Both red-bellied snakes and copper-bellied water snakes lack
well-defined black dots at the outer edge of each ventral scute.

American Water Snakes, genus *Nerodia*

The American water snakes are, for the most part, rather large snakes that, if carelessly restrained, can vary from moderately defensive to downright irascible. When handled these snakes often bite or smear an unpleasant combination of musk, feces, and urates on their captor.

As with many snakes possessing a Duvernoy's gland, the saliva of water snakes contains complex proteins that may prove mildly toxic to some persons.

All of the American water snakes are live bearing, some having immense litters. Recently a large female Florida green water snake, just under 5 feet and killed on the road, was found to contain 128+ well-developed young.

With 9 species (a total of 20 subspecies), eastern North America is generously endowed with water snakes. These vary in size from the hulking Florida green water snake to the comparatively diminutive Brazos water snake. No matter the size, all are comparatively heavy bodied, and heavily gravid females may look almost grotesquely overweight.

Sadly, many of these snakes (especially those with darker color and minimal pattern) are regularly mistaken for the venomous cottonmouth and killed. Water snakes are often seen basking by day, even in fairly cold but sunny weather. They often choose warmed concrete abutments, protruding rocks or snags, or limbs overhanging the water as basking sites. Basking water snakes are wary and will often drop into the water and dive at the first sign of disturbance. They may surface rather quickly and either scull slowly in place or swim parallel with the shore to assess the severity of the disturbance. If again frightened they often submerge and remain below the surface for long periods. None of our water snakes stand their ground and gape at an intruder.

Water snakes may be very active on warm, rainy or humid spring and summer nights. During such times of peak activity, water snakes often cross roadways, where carnage exacted by vehicles can take a terrible toll.

Water snakes have heavily keeled, rather dull scales and, in nearly all instances, a divided anal plate. However, the several races of the red-bellied water snake may occasionally have undivided anal plates.

The snakes of this genus are almost always found near water. They are often seen by boaters on quiet, secluded waterways, and may be particularly common on flooded prairies and along canals.

103. Gulf Salt Marsh Snake

Nerodia clarkii clarkii

Disposition: All salt marsh snakes may bite when disturbed but are harmless to humans.

Abundance/Range: This is a common but wary and secretive, hence seldom-seen, snake. The Gulf salt marsh snake ranges eastward from just south of San Antonio Bay, Texas, to Florida's Cedar Key.

Habitat: The salt marsh snakes (collectively) are at home in a habitat used by comparatively few reptiles. Look for them in tidally influenced brackish and saltwater habitats, especially salt marshes.

Size: Although some may near 36 inches in length, most Gulf salt marsh snakes are adult at 18–30 inches. Neonates average 9 inches in length.

Identifying features: Unusual in a genus of primarily blotched or banded snakes, the Gulf salt marsh snake is prominently striped. There is very little ontogenetic (age-related) change from neonate to full adult. The dorsal ground color is gray to olive. There are 4 wide, dark stripes (2 dorsolateral and 2 lateral) running from nape to tail. The belly is reversed in color from that of the dorsum, being reddish with a central light line. The heavily keeled scales are in either 21 or 23 rows and the anal plate is divided.

Similar snakes: Use range to help identify the Gulf salt marsh snake. The various striped crayfish snakes have 19 rows of scales. Garter and ribbon snakes lack the midventral spots. Patch-nosed snakes have smooth scales.

Comments: As high tides flood the salt marshes these snakes disperse, returning to the tidal creeks as the tides wane. The snakes bask and forage during day-

103. Gulf Salt Marsh Snake

Gulf
mangrove
Atlantic

light hours in cool weather, but become largely nocturnal during the hot summer nights.

Salt marsh snakes often forage for fish trapped in tiny tide pools. The salt marsh snakes may seek refuge in the burrows of the crabs or crayfish that abound in salt marsh habitat. Many of the snakes have imperfect tails, perhaps the result of injuries and amputations by clawed crustaceans.

Additional subspecies

104. The Mangrove Salt Marsh Snake, *Nerodia clarkii compressicauda*, is the most variably colored of the three races of this snake. Some individuals may be strongly patterned and others essentially unicolored. To complicate identification even more, the mangrove salt marsh snake intergrades extensively with the Gulf salt marsh snake on Florida's Gulf coast and with the Atlantic salt marsh snake on the Atlantic coast.

The ground color of the mangrove salt marsh snake may be tan, rich russet, black, gray, or olive. The snakes with the darker ground colors are often blotched or banded with darker color. Occasionally they may be striped anteriorly and blotched posteriorly. The reddish and tan examples are banded when young, may retain vestiges of the bands as they grow, but often become

104. Mangrove Salt Marsh Snake

unicolored adults. The belly is often quite like the dorsum in color, but paler and usually lacks a well-defined pattern. The scales are heavily keeled, in either 21 or 23 rows, and the anal plate is divided. The normal size of the mangrove salt marsh snake is 24–36 inches. This subspecies ranges in brackish and saltwater habitats from Tampa Bay south to the Lower Keys, then northward on Florida's east coast to Cape Canaveral. This race has 6–20+ young.

105. The Atlantic Salt Marsh Snake, *Nerodia clarkii taeniata*, is a federally endangered subspecies. At a record size of just under 24 inches, it is the smallest of the three races. The Atlantic salt marsh snake is striped anteriorly and banded or blotched posteriorly. The dorsal ground color is gray to olive. The 4 anterior stripes (2 dorsolateral and 2 lateral) may be barely darker than the ground color. The posterior lateral blotches are often the most prominent markings. The belly scutes are reddish and each bears a yellowish midventral spot. The heavily keeled scales are in either 21 (normal) or 23 rows and the anal plate is divided.

This rare snake may be found in its pure form only in a few tidal waterways of

105. Atlantic Salt Marsh Snake. Photo by R. Wayne Van Devender

Volusia County, Florida. It is known to intergrade with the mangrove salt marsh snake where the ranges abut.

106. Mississippi Green Water Snake

Nerodia cyclopion

Disposition: This is an irascible snake species that will bite repeatedly when restrained.

Abundance/Range: This snake is common from Texas to Illinois and Alabama. It barely enters Florida in the vicinity of Escambia Bay.

Habitat: Open marshlands, bayous, rivers, canals, rice fields, creeks, and the weedy edges of ponds and lakes are all suitable habitats for this common water snake.

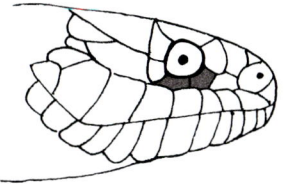

Figure 5.05. Mississippi Green Water Snake and Florida Green Water Snake, showing suboculars. Illustration by Patricia Bartlett.

106. Mississippi Green Water Snake

Size: This is the smaller of the two species of green water snakes. Adults vary in size at 20–30 inches, with occasional females (the larger sex) attaining 48 inches. Neonates measure about 7 inches in length.

Identifying features: Adult Mississippi green water snakes are often less contrastingly patterned than juveniles, but this snake does not undergo particularly extensive ontogenetic changes. Mud will often adhere to the scales, obscuring the pattern. Pattern is most visible when the snake is wet or freshly shed.

The dorsal ground color is olive brown to olive green. There is usually at least a vestige of darker bands; these are best defined on the sides. The belly is yellowish to olive with an often well-defined pattern of lighter spots or crescents. There are subocular scales between the bottom of the eye and the upper lip (supralabial) scales. The chin is often yellowish. The Mississippi green water snake has either 27 or 29 rows of strongly keeled scales. The anal scale is divided.

Similar snakes: The Florida green water snake lacks a belly pattern. Other water snakes (except the Florida green) lack the subocular scales. The cottonmouth has a strong facial pattern.

107. Red-bellied Water Snake

Nerodia erythrogaster erythrogaster

Disposition: When hard pressed, these big natricines can be spectacularly feisty. They strike hard and bite readily. They should be handled carefully.

Abundance/Range: This snake may be abundant in suitable habitats. It ranges northward from north Florida and southeastern Alabama, through much of Georgia, to the central Delmarva Peninsula.

Habitat/Range: These beautiful water snakes inhabit a multitude of aquatic

red-bellied
yellow-bellied
copper-bellied
blotched

107a. Red-bellied Water Snake, adult

107b. Red-bellied Water Snake, juvenile

habitats, including springs, river edges, cypress domes, and open marshes. Ponds, sloughs, swamps, and oxbows are also favored. They are occasionally encountered far from the nearest permanent water.

Size: Although females (the larger sex) often exceed 42 inches in length (rarely, they may attain a length of 60 inches), they are usually somewhat smaller. Neonates average 10 inches in length.

Identifying features: Considerable ontogenetic changes occur in this water snake. Neonates are prominently blotched or irregularly banded with blackish brown on a gray ground color. The chin is yellowish. The belly is pale. As they mature the dorsal and lateral patterns become pale. Adult snakes are nearly unicolored dark brown to blackish dorsally and have a bright orange-red belly. The heavily keeled scales are in 19–23 rows at midbody. A great majority (but not all) of the specimens have divided anal plates.

Mud and dust may adhere to the scales, obliterating the true colors.

Where the ranges of the red-bellied water snake and the yellow-bellied water snake abut, intergradation occurs. The intergrades often have a yellowish red venter.

Similar snakes: Swamp snakes are much smaller and have smooth scales. Red-bellied snakes are small and have 3 light spots behind the head. These may coalesce into a pale collar. Northern water snakes and queen snakes have a patterned belly with a pale ground color. Kirtland's snakes have a blotched dorsum and black spots bordering the red of the belly. The more westerly yellow-bellied water snake is very similar but has a yellow (not red) belly.

Comments: This water snake species occasionally wanders far from water. It is, however, most commonly seen swimming near springs or basking on riverbanks.

Additional subspecies

108. The Yellow-bellied Water Snake, *Nerodia erythrogaster flavigaster*, also undergoes dramatic ontogenetic color changes. Neonates are prominently blotched or irregularly banded with blackish brown on an olive gray ground color. The chin is yellowish. The belly is pale. As the snake matures the dorsal and lateral

108. Yellow-bellied Water Snake

patterns pale and with adulthood it assumes a nearly unicolored olive black dorsum and a creamy yellow to bright yellow belly. Mud and dust may conceal the snake's color. The heavily keeled scales are in 19–23 rows at midbody. A great majority (but not all) of the specimens have divided anal plates. The record size for this race is 59 inches.

This snake may be found from Florida's western panhandle to central Georgia, Illinois, and eastern Texas.

109. The Copper-bellied Water Snake, *Nerodia erythrogaster neglecta*, is an uncommon inland subspecies. It is usually somewhat smaller than the other races, being adult at 25–40 inches.

There is evidence that the copper-bellied water snake inhabits ephemeral waters in the spring and moves to permanent water sources as the shallow ponds dry up. The snakes bask and forage during daylight hours in the spring, increasing their hours of activity as temperatures warm with the long days of summer. They inhabit ponds, sloughs, swamps, oxbows, and river edges, in or near deciduous woodlands. Occasionally they may be seen in more open areas. Copper-bellied water snakes have a curiously disjunct range. They were once probably more generally distributed, but postglacial climatic and topographical changes, as well as human-caused habitat degradation, have concentrated these snakes in tiny pockets of suitable habitat. They are found in northwestern Tennessee, northwestern Kentucky and adjacent Indiana and Illinois, south central

109. Copper-bellied Water Snake

Indiana, western central Ohio, south central Michigan, and adjacent Indiana and Ohio.

Neonates are prominently blotched or irregularly banded with brown on a russet to gray ground color. The belly is pale. As they mature the contrasts fade and with adulthood the snake assumes a nearly unicolored dark brown to blackish dorsum and a yellowish to coppery venter. The venter may be most intensely colored along the midline. On many specimens the dark dorsal coloration extends to and at least dusts the edges of the ventral scutes. The scales are heavily keeled and in 19–23 rows at midbody.

110. The Blotched Water Snake, *Nerodia erythrogaster transversa*, is the least colorful and most western race of this species. It ranges southward from northwestern Kansas and adjacent Missouri through Texas to north central Mexico. It inhabits almost any type of water-retaining habitat, from stock watering tanks to river edges. The dorsal color varies from brown to slate gray, with a series of variably distinct dark-edged dorsal blotches that are darker than the ground color. Smaller lateral blotches are also present. Old snakes are often less well marked than young adults. Those from the western portions of the range are often of lighter color than those from the east. The belly of adults is yellow, often with a blush of orange along the edges. Juveniles are very prominently patterned.

110. Blotched Water Snake

111. Banded Water Snake

Nerodia fasciata fasciata

Disposition: This snake bites readily if carelessly restrained.

Abundance/Range: This is an abundant water snake—perhaps *the* most abundant water snake—throughout its range. This race ranges northward from the Florida panhandle, adjacent Alabama, and southern Georgia along the Coastal Plain to northern North Carolina.

Habitat: The banded water snake occupies virtually every aquatic habitat throughout its range except pure saltwater.

Size: This snake is adult at 24–40 inches, but may occasionally attain a full, heavy-bodied 60 inches in length. Neonates average about 9½ inches overall.

Identifying features: Both the ground color and the pattern of the banded water snake are variable. The ground color may be gray, tan, olive, or almost black. The pattern may be dark olive, greenish brown, old rose, or red. The bands (this snake is banded for its entire length) may be outlined with darker pigment. There are usually no dark lateral spots between the dark bands. Some banded water snakes are so dark that they may appear almost unicolored. If light areas are present they will be on the lower sides. The belly is off-white to white and has an irregular pattern of darker (sometimes a quite bright red) pigment. The upper labial scales are usually outlined by dark pigment. A dark stripe may run from the back of the eye to the rear of the angle of the jaw. Neonates are more strongly patterned than most adults.

Similar snakes: The several races of the northern water snake can be of very

111. Banded Water Snake

banded
broad-banded
Florida

similar appearance to the banded water snake. However, the posterior markings of the northern water snake are blotches rather than bands.

Comments: Collectively, the three races of *N. fasciata* are referred to as the southern water snakes. Warm springtime evening rains will often prompt vast numbers of these snakes to disperse. Many are males, seemingly following pheromone trails of sexually receptive females.

Additional subspecies

112. The Broad-banded Water Snake, *Nerodia fasciata confluens*, is adult at a length of 24–40 inches. The record size is 45 inches. It ranges from eastern Texas and extreme southern Illinois to southeastern Louisiana. Intergrades between this species and the more easterly banded water snake occur in southern Mississippi and southern Alabama.

Like that of the banded water snake, the ground color and pattern of this subspecies are variable. The ground color may be gray, tan, olive, or almost black. The pattern may be dark olive, greenish brown, old rose, or red. The bands are often the darkest at their outer edges. There are usually no dark lateral spots between the dark bands, but there may be light areas in the dark bands. Some of these water snakes are so dark that they may appear almost unicolored. The belly is off-white to white and bears a pattern of comparatively large, darker

112. Broad-banded Water Snake

(sometimes quite bright red) squared markings. The upper labial scales are usually outlined in their vertical sutures by dark pigment. A dark stripe may run from the back of the eye to the angle of the jaw. Neonates are more strongly patterned than most adults.

113. The Florida Water Snake, *Nerodia fasciata pictiventris*, is the southernmost representative of this species. It is adult at 24–42 inches, but rare individuals may attain 60 inches in length. An abundant snake, it is restricted in distribution to the Florida peninsula and the adjacent Okefenokee swamp in Georgia.

113. Florida Water Snake

The ground color may be gray, tan, olive, or almost black. The pattern may be dark olive, greenish brown, old rose, or red. The bands may be outlined with darker pigment. There may be dark spots on each side between the dark bands. Some snakes are so dark that they may appear almost unicolored. If light areas are present they will be on the lower sides. The belly is off-white to white and has an irregular pattern of darker (sometimes a quite bright red) pigment. The upper labial scales are usually outlined by dark pigment. A dark stripe may run from the back of the eye to the rear of the angle of the jaw. Neonates are more strongly patterned than most adults.

114. Florida Green Water Snake

Nerodia floridana

Disposition: Big and grouchy, the Florida green water snake is always ready to bite—and to bite hard—if carelessly restrained.

Abundance/Range: When water levels are normal, this snake disperses widely and is commonly seen throughout most of its range. However, during years of drought, it is found only in the immediate proximity of standing permanent water. It is found throughout most of Florida and northward along coastal Georgia to southeastern South Carolina.

Habitat: This big snake can be abundant along rivers, canals, creeks, and flooded prairies and is sometimes found in ponds and lakes. It prefers unshaded waters.

Size: This is the larger of the two species of green water snakes. Adults measure 3–4 feet in total length, with occasional examples exceeding 6 feet long. Neonates are about 9 inches long.

Identifying features: This is a large and dull-colored water snake. Adults are less contrastingly patterned than juveniles, and mud often obliterates their true

114a. Florida Green Water Snake, normal morph

114b. Florida Green Water Snake, reddish morph

colors. To see any pattern on an adult snake, it is often necessary to see the snake when it is either wet or freshly shed.

The dorsal ground color is olive brown to olive green. There is usually at least a vestige of darker bands. The belly is yellowish to olive and unpatterned except beneath the tail. There are several scales (the suboculars) between the bottom of the eye and the upper lip scales (supralabials). This character is shared with the closely related Mississippi green water snake. The chin is yellowish green. The Florida green water snake has either 27 or 29 rows of strongly keeled scales. The anal scale is divided.

Similar snakes: The Mississippi green water snake has a prominent belly pattern. Other water snakes (except the Mississippi green) lack the subocular scales. The cottonmouth has a strong facial pattern.

115. Brazos Water Snake

Nerodia harteri harteri

Disposition: This is a small water snake that may bite if carelessly restrained.

Abundance/Range: Because of habitat modifications by humans, this has become a rare snake that is now protected by law in the state of Texas. This small water snake occurs only in the Brazos River drainage of north central Texas.

Habitat: Primarily adapted to life among rocks in rather swift riffles, this snake

115. Brazos Water Snake

Brazos
concho

now also inhabits the rocky shorelines of a few lakes (such as Lake Granbury and Possum Kingdom Lake) that have been created by impounding the Brazos River.

Size: This snake is relatively slender for a water snake. Adults are often 20–25 inches long, but seldom more than 30 inches. The record size is 35½ inches. Neonates are 7–10 inches long.

Identifying features: Certainly not brightly colored, this rare little water snake nonetheless has an attractive ground color of sandy tan, olive tan, or light olive brown. A lighter vertebral stripe is often present. This snake has 4 rows of olive spots along its length. The 2 uppermost rows are the largest and best defined. If no vertebral stripe is present, these spots may be roughly conjoined vertebrally. If the vertebral stripe is present it separates these 2 rows of spots. The spots in the lower row along each side alternate between the spots in the dorsolateral rows. The belly is cream, pale yellow, or greenish yellow anteriorly and shades to pale pink posteriorly. Each outer edge of each ventral plate bears a black dot. The posterior pair of chin shields are usually separated from each other by 2 rows of small scales. Juveniles are similar to the adults in appearance but reportedly have more contrasting dorsal patterns and a brighter belly color.

Similar snakes: The Texas brown snake is smaller, has dark nape spots but lacks the 4 rows of body spots, and has a whitish venter. The blotched water snake is usually grayish above and has very large dorsal saddles rather than the 2 rows of spots. See the next account (number 116) for comments on the related Concho water snake.

Comments: If quiet approach is made you might be lucky enough to see a Brazos water snake angling for small fish. It holds its position by anchoring its tail in the rocks. Neonates seem to prefer shallow riffle edges or side pools; adults are most often found in deeper water. Juveniles often seek shelter beneath riverside flat rocks.

The normally stated subspecific differences actually overlap and may be inconclusive. We suggest that for the moment you differentiate these snakes by range rather than by appearance.

Additional subspecies

116. The Concho Water Snake, *Nerodia harteri paucimaculata*, has derived its subspecific name from a supposed paucity of dark spots on the ventral plates. We have found this to be a variable characteristic, some examples having virtually no spotting, some having a few spots, and others having many. The body coloration of those we have seen has been a definite buff with 4 rows of variably distinct blotches. The belly is yellowish anteriorly and pinkish posteriorly. The posterior pair of chin shields seldom have more than 1 row of small scales separating them, and often have none. Juveniles are somewhat more richly colored. This snake occurs along the Upper Concho and Colorado River drainages in Texa. It is also found in Ivie Reservoir, E. V. Spence Reservoir, and Ballinger Municipal Lake. This rare snake is listed as an endangered species by both fed-

116. Concho Water Snake

eral and state decree. The record size is 41¾ inches, but most examples found are in the 20–30–inch range.

117. Diamond-backed Water Snake

Nerodia rhombifer rhombifer

Disposition: This snake will bite readily if carelessly restrained.

Abundance/Range: This is an abundant snake throughout most of its range, westward and northward from central eastern Alabama to central Texas and southeastern Iowa. Its range in the north closely follows river systems.

Habitat: Rivers, ditches, bayous, swamps, marshes, lakes, ponds, rice fields, cattle tanks, and virtually any other body of water is utilized by this snake in the south. In the north it is more or less restricted to low elevations and riverine habitats. We have seen diamond-backed water snakes in the clean waters of irrigation canals and lakes and in waters badly fouled by dead and decomposing fish and amphibians.

Size: Females often attain or even exceed a heavy-bodied 54 inches in length. Males tend to be considerably smaller, but may occasionally grow nearly as large as the females. The record size is in dispute, but nears or slightly exceeds 5½ feet. Neonates are 9–12½ inches in length.

Identifying features: This can be a nondescript snake, especially prior to shedding or when its scales are coated with dry mud. It has a ground color of olive brown, olive green, light brown, orange brown, or yellow green. A pattern of dorsal diamonds is outlined in black along most of the back; a short distance anterior to the vent and on the tail the dark markings tend to become bars,

117. Diamond-backed Water Snake

and numerous dark, broad, vertical bars are present on each side. One tends to connect with the lowest point on each dorsal diamond. The head and face are usually devoid of dark markings. The eyes are intense copper. Adult males may have papillae (tubercles) on the chin. The belly is yellowish. The outer edge of each belly scute (scale) usually bears an irregular dark marking. Additional smaller dark marking may be present centrally on some scutes.

Similar species: This big, dark, sullen-appearing aquatic snake is often mistaken for the venomous cottonmouth. Cottonmouths have elliptical pupils, heat-sensory facial pits, and horizontal facial patterns that the diamond-backed water snake does not have. The Mississippi green water snake is of similar robust build but has almost no discernible pattern on its dorsum and has a row of small scales (the suboculars) between the bottom of the eye and the lip scales.

Comments: We have seen large numbers (10–15 in some cases) of males, little more than half the length and less than half the diameter of what was apparently a sexually receptive basking female stretched on the bank of a rice paddy, crawling actively about and over her. Several of the males were intertwined and "wrestling" for dominance.

118. Northern Water Snake

Nerodia sipedon sipedon

Disposition: Although juvenile northern water snakes will often allow themselves to be handled without biting, this is not usually the case with adults.

Abundance/Range: This is one of the most common snakes in eastern North America. It ranges southward from eastern Maine and southeastern Ontario to the mountains of western North Carolina and eastern Tennessee.

Habitat: This is a wide-ranging snake that occupies pond and lake shores, river and stream edges (including areas with a fair current), marsh and swamp margins, oxbows, and, often, temporarily flooded areas.

Size: Adults vary in size between 24 and 54 inches, with females being the larger sex. Neonates measure about 9¼ inches at birth.

Identifying features: Although adults are often less brightly colored than juveniles, the northern water snake does not undergo particularly extensive ontogenetic changes. Mud will often adhere to the keeled scales, making these snakes appear even less colorful than they actually are. To fully appreciate the brilliance and intricacy of the colors and patterns, it is often necessary to see the snake when it is either wet or, better yet, freshly shed.

118. Northern Water Snake

In the northeastern states and southern Canada, range alone will provide the identity of this species. However, where the range abuts or overlaps that of other species of water snakes, confusion may prevail.

The ground color of the northern water snake may be gray, buff, tan, or brown. The markings are usually much darker than the ground color, but this is not invariably so. The markings of the juveniles are often blackish, as may be those of some adults. On other examples the markings may be reddish, with or without a black border. Typically the anterior markings are in the form of rather regular, broad bands. At a point somewhat anterior to midbody, the banding becomes irregular or may change to a combination of dorsal saddles and lateral blotches (occasional specimens are regularly banded for their entire length). Tail bands may be regular or irregular. The belly is usually strongly patterned, but may be sparsely so, or virtually unpatterned except for a sparse to liberal dusting of dark dots. If markings are present, they may be large and in the shape of dark-bordered half moons or irregular triangles, apex directed posteriorly. They are usually arranged on each side of a complete or broken dark-edged light midline, and may be present only at the very edges of the ventral scutes. Any given specimen may combine any or all of these colors and patterns. Northern water snakes usually lack a horizontal dark ocular (eye) streak and have 21–25 scale rows and a divided anal scute.

Similar snakes: Although banded water snakes are equally variable in color, they usually have rather regular bands along their entire length and a dark, horizontal eye streak. When adult the various races of the red-bellied water snake have unmarked bellies.

Additional subspecies

119. The Lake Erie Water Snake, *Nerodia sipedon insularum*, is restricted in distribution to shores of islands of Ohio's Put-in-Bay Archipelago near the southwestern tip of Lake Erie.

119. Lake Erie Water Snake

On a recent visit to its habitat we found this water snake to be quite common on many of the rocky beaches overlooked by low cliffs and along jetties.

The Lake Erie water snake is of lighter ground color and less contrastingly patterned than the northern water snake. It is often gray to gray green dorsally with reduced or absent dorsal and ventral patterns. Interbreeding with the darker water snakes of the mainland has, in many cases, produced darker than normal ground colors and rather strong patterns. Still, researchers concede that the water snakes of Put-in-Bay are, on average, lighter and less contrasting in color than those from the mainland. The ventral scutes have dark pigment along the sides and the anterior edge. At an adult size of 30–40 inches, females are the larger sex.

This snake may change color with ambient temperature, being lighter when hot. Neonates are more strongly patterned than most adults.

120. The Midland Water Snake, *Nerodia sipedon pleuralis*, ranges westward and southward from central southern North Carolina to eastern Missouri and Mississippi and extreme western Florida. This abundant snake basks on all manner of haul-outs.

Adults are often less brightly colored than the similarly patterned juveniles. The ground color of the midland water snake may be gray, buff, tan, brown, or

120. Midland Water Snake

reddish. The markings are usually much darker than the ground color, but this is not invariably so. The markings of the juveniles are often blackish, as may be those of some adults. On other examples the markings may be reddish, with or without a darker border. Typically the anterior markings are in the form of rather regular, broad bands. At a point somewhat anterior to midbody, the banding becomes irregular or may change to a combination of dorsal saddles and lateral blotches (occasional specimens are regularly banded for their entire length). Tail bands may be regular or irregular. The belly is usually strongly patterned but may be sparsely so, or virtually unpatterned except for a sparse to liberal dusting of dark dots. If markings are present, they are often in the form of 2 rows of irregular triangles, apex directed posteriorly. The color of the dark bands often encroaches onto the outer edges of the belly scales. The midland water snake usually lacks a horizontal dark ocular (eye) streak and has keeled scales in 21–25 scale rows and a divided anal scute. The record size is 59 inches.

121. The Carolina Water Snake, *Nerodia sipedon williamengelsi*, is the darkest of the races of northern water snakes and has the most restricted range. It is confined in distribution to the shoreline of central eastern mainland North Carolina and to the Outer Banks. It seems most common where salt grasses such as *Spartina* and *Juncus* grow thickly along canals, ditches, salt marshes, creeks, and interior freshwater ponds. Despite continuing controversy regarding its taxonomic validity, this snake is specifically listed by the state of North Carolina as a species of special concern.

121. Carolina Water (salt marsh) Snake

Neonates are prominently banded with dark against a variable gray ground color. As the Carolina water snake grows, a suffusion of melanin darkens its dorsum (darkest middorsally), occasionally obscuring most of the pattern. The lighter ground color separating the blotches is usually no more than 1½–2 (rarely 3) scales wide. Although the ventral blotches anterior to the 50th ventral scute may have tan or brown centers, those posterior to that scute are entirely black. At 40–55 inches, females seem, at least marginally, to be the larger sex.

122. Brown Water Snake

Nerodia taxispilota

Disposition: Succinctly stated, this snake is big and feisty, ready to bite if carelessly restrained, and a bite by a large example is unpleasant.

Abundance/Range: The brown water snake is commonly seen when water levels are normally high, but may be localized and difficult to find when water levels have been reduced by drought. The brown water snake is found throughout all of mainland Florida as well as southern Alabama and most of Georgia, and northward along the coastal plain to southern Virginia.

Habitat: This snake can be abundant along rivers, canals, creeks, and flooded prairies and is sometimes found in ponds and lakes. It prefers habitats with ample dry, sunny basking sites such as logs or shrubs projecting over the water, or rock piles or grassy banks along the water's edge.

Size: When the immense girth of a gravid female is considered, the brown water snake is among the largest snakes of eastern North America. Adult females

122. Brown Water Snake

may attain 66 inches in length. Males are usually substantially smaller than the adult females. Neonates are quite variable in size, measuring about 8–12 inches in length at birth.

Identifying features: Although these snakes usually bear a rather well-defined pattern, this may be nearly obliterated if the scales are muddy. Patterns are very easily seen on wet or freshly shed snakes. Juveniles are the most contrastingly patterned and brightly colored.

The dorsal ground color is dark tan to dark brown. A series of large, dark dorsal blotches is typical. Dark lateral spots, somewhat smaller than the dorsal markings, alternate with the latter from neck to tail. The lateral spots do not touch the dorsal blotches. The belly is lighter than the dorsum. Each ventral scute bears 2 or 3 brown spots that form irregular lines. The chin is creamy brown to light brown. The strongly keeled scales of the brown water snake may vary in row count between 25 and 33. The anal scale is divided.

Similar snakes: This is the water snake of the eastern seaboard that is most commonly mistaken for the cottonmouth. However, although its head is much wider than its neck, the brown water snake lacks the strong facial pattern so typical of a cottonmouth and does not gape when disturbed. The diamond-

backed water snake can look very similar, especially when patterns are obscured by mud. However the ranges of these two snakes barely abut, and the bars on the side of the diamond-backed water snake contact the dorsal markings.

Crayfish Snakes and Queen Snakes, genus *Regina*

The four species in this genus are aquatic and prey specialists, preferentially feeding on crayfish (which are usually ingested tail first). Two, the queen snake and the Graham's crayfish snake, seem to prefer soft-shelled (freshly molted) crayfish while the other species are less discriminating, opportunistically eating both hard- or soft-shelled crayfish. Recent studies have disclosed that some component of the saliva of the glossy crayfish snake is toxic to crayfish, quickly rendering the crustaceans immobile and easily swallowed. Aquatic worms and fish may occasionally be eaten.

All members of this genus are live bearing. Heavily gravid females attain a proportionately immense girth. The litter size usually consists of fewer than a dozen 6–8-inch neonates, but the Graham's crayfish snake is known to have had 39. Parturition occurs in mid to late summer.

Although a held crayfish snake may twist and thrash in an effort to free itself, it will usually not attempt to bite. It may, however, smear musk and feces on its captor.

These snakes may leave their ponds, sloughs, and rivers and move overland on warm, rainy spring and summer evenings. They may also be found by raking floating aquatic vegetation on to the shore and sorting through it or by turning waterside debris (including rocks and mats of vegetation). The queen snake and Graham's crayfish snake often bask on limbs or bushes overhanging the water.

The dorsal scales of all four species are in 19 rows at midbody and the anal plate is divided. Three of the four species have keeled scales with only the striped crayfish snake having smooth scales. The striped and the glossy crayfish snakes are stout and shiny; the Graham's crayfish snake and the queen snake are rather slender, nonshiny species.

This genus is restricted in distribution to the eastern United States.

123. Striped Crayfish Snake

Regina alleni

Disposition: Although the saliva of this species seems to contain an agent that is toxic to crayfish, it is not considered dangerous to humans. These snakes rarely bite when captured.

Abundance/Range: Because of the secretive habits of this snake, its actual abundance is difficult to assess. A search of favorable habitats, by day or night, will usually bring a specimen or two to light. Dozens to hundreds of these snakes may be seen crossing canal-edge roadways on warm, rainy spring nights. This snake ranges over all of peninsular Florida, the eastern half of the Florida pan-

123. Striped Crayfish Snake

handle, and adjacent south central Georgia.

Habitat: Look for this little snake among mats of water lettuce, hyacinth, and pennywort and in tangles of other aquatic plants at the edges of slow creeks, rivers, lakes, ponds, swamps, canals, cypress heads, and flooded ditches.

Size: The striped crayfish snake is adult at a stout 15–20 inches. Occasional examples may attain 26 inches in length.

Identifying features: The shiny scales of the striped crayfish snake are precisely colored. The back and upper sides are olive brown to dark brown. A dark dorsolateral stripe one scale in width is present on each side, as is a vertebral stripe of similar width. The lower sides and belly are yellow, often with olive overtones. Two or three narrow dark stripes are present in the yellow field of the lower sides. The belly is often unmarked, but may have a midventral row of small black dots. The scales are feebly keeled above the vent and on the tail; the smooth scales are in 19 rows. The anal plate is divided. The head is narrow and the supralabials are yellow.

Similar snakes: Garter snakes, ribbon snakes, and the other species of crayfish snakes all have keeled dorsal scales.

124. Graham's Crayfish Snake

Regina grahamii

Disposition: Virtually nothing is known about the salivary enzymes of Graham's crayfish snake but this little snake almost never attempts to bite humans.

Abundance/Range: Because of the secretive habits of this snake, its abundance can be difficult to assess. It seems to fare reasonably well despite rampant encroachment on its habitats. Fair numbers have been found crossing roadways adjacent to flooded rice fields, swamps, bayous, or marshes on rainy spring and summer nights. The Graham's crayfish snake ranges northward from eastern Louisiana and central Texas to central Iowa.

Habitat: Streams, rice fields, muddy roadside ditches, wet prairies, bayous, swamps, and their immediate environs are the habitat of this snake. It may occasionally be found sunning on the banks or in shrubs or low in trees that overhang the water. When out of the water, this snake is very secretive, seeking seclusion beneath logs, stones, boards, or mats of vegetation.

124. Graham's Crayfish Snake. Photo by Terry Hibbitts

Size: With a record size of 47 inches, this is by far the largest of the four species in this genus. A more normal adult size, however, is 20–30 inches. Neonates measure 7–10 inches in length.

Identifying features: The very aquatic Graham's crayfish snake has a ground color of olive gray or olive tan to a very dark olive brown dorsally (some brown examples lack all traces of pattern). It is predominantly yellow to pale brown ventrally. When a pattern is present it usually consists of a wide olive yellow lateral stripe on each side (scale rows 1, 2, and 3). The yellow stripe is separated from the olive gray ground color dorsally by a straight, dark stripe. It is separated from the belly scales by a dark zigzag stripe. A narrow, dark midventral stripe is present as is a lighter vertebral stripe. The lips are cream to yellow, the throat is white to off-white.

The 19 rows of scales are keeled. The anal plate is divided.

Similar snakes: Glossy crayfish snakes have shiny scales. Garter snakes are usually prominently striped and have undivided anal plates. The Gulf salt marsh snake has 4 dark dorsal stripes.

125. Glossy Crayfish Snake

Regina rigida rigida

Disposition: The saliva of this species contains elements toxic to crayfish, but the snake seldom bites humans. The teeth of the glossy crayfish snake, which have been called "stout and chisel-like," are hinged at the base.

Abundance: This is a secretive but rather abundant snake. A search of favorable habitats will usually disclose a specimen or two. These snakes often cross roadways adjacent to swamps, marshes, or other bodies of water on rainy spring and summer nights. This race occurs in the northern one-third of the Florida peninsula, the eastern half of the panhandle, and from there northward along

125. Glossy Crayfish Snake

the coastal plain to northern North Carolina. A disjunct population occurs in central eastern Virginia.

Habitat: Streams, rivers, and their immediate environs are most favored by this snake, but it can also be found in lakes, ponds, swamps, canals, and flooded ditches. In waterside habitats, glossy crayfish snakes seclude themselves beneath logs, boards, or mats of vegetation. They also seek refuge in floating mats of hyacinths, water lettuce, and pennywort.

Size: The glossy crayfish snake is adult at a stout 18–25 inches. Occasional examples may reach a length of 2½ feet. Neonates are 7–8 inches long.

Identifying features: The three races of the glossy crayfish snake are difficult to differentiate. Where in the overall range of the species your example was found will be a useful identification tool.

The dorsal coloration is a shiny dark olive brown to brown, often with thin dark lines at the edge of the ventral scutes and paler dorsolateral stripes weakly visible. The sides of the throat bear thin dark stripes. Placing the snake in water may make the striping more easily visible. The scales of the upper lip are yellow.

The belly is yellowish (darkest on old snakes) and bears 2 rows of bold, similarly sized semicircles.

The shiny scales are keeled and in 19 rows. The anal plate is divided. Females of this race usually have fewer than 55 rows of subcaudal scales (the scales beneath the tail); males have fewer than 63 rows of subcaudals.

Similar snakes: Graham's crayfish snake and the queen snake do not have shiny scales. Garter snakes are usually prominently striped and have undivided anal plates and dull scales. The striped crayfish snake is visibly striped.

Additional subspecies

126. The Delta Crayfish Snake, *Regina rigida deltae*, is found in eastern Louisiana and adjacent southwestern Mississippi. It is a weakly defined race, differentiated only by the lack of dark lines on the throat and the fact that it usually has only one prefrontal scale (at least on one side of the face) rather than two. The belly is yellowish (darkest on old snakes) and bears 2 rows of bold, regularly sized semicircles. The scales are keeled and in 19 rows.

126. Delta Crayfish Snake. Photo by R. Wayne Van Devender

127. The Gulf Crayfish Snake, *Regina rigida sinicola*, is found both to the west and to the east of the Delta crayfish snake. This race occurs from eastern Texas to central Louisiana and again from central southern Mississippi to southeastern Georgia. Depend largely on range for identification. The dorsal coloration is a shiny dark olive brown to brown. Light dorsolateral and dark ventrolateral stripes are weakly visible. Placing the snake in water may make the striping more easily visible. There are no dark stripes on the sides of the throat. It has 2 prefrontal scales on each side of the face. The scales of the upper lip are yellow. The belly is yellowish and bears the usual 2 rows of bold, regularly sized semicircles.

127. Gulf Crayfish Snake

The scales are keeled and in 19 rows. Females have more than 54 rows of subcaudal scales; males have more than 62 rows.

128. Queen Snake

Regina septemvittata

Disposition: This snake produces salivary enzymes that are toxic to crayfish but harmless to humans.

Abundance/Range: This is a secretive but not uncommon snake that can occasionally be found crossing roadways during rainy weather. Queen snakes prefer to dwell along streams and small rivers but are also found in swamps, marshes, or other bodies of water.

Avoiding the coastal plain, the snake may be found on Florida's western panhandle and north and west to Pennsylvania, Wisconsin, and Mississippi. A large disjunct population occurs in northwestern Arkansas and immediately adjacent Missouri.

Habitat: Streams, rivers, and their environs are most favored by this snake. North of Florida it is associated with the rocky riparian areas, but in Florida—a largely rockless state—the queen snake occurs in cypress heads and along creeks and small rivers. It may often be seen sunning on shrubs or low in trees overhanging the water. When out of the water, the queen snake is very secretive, seeking seclusion beneath logs, boards, or mats of vegetation.

128. Queen Snake

Size: Although often smaller, the queen snake can attain a slender 36 inches when adult. Neonates measure 7–9 inches in length.

Identifying features: The very aquatic queen snake is strongly bicolored. It varies from olive tan to a very dark olive brown dorsally and is yellow(ish) to russet ventrally. There is a wide cream to yellow lateral stripe on each side. The upper lips are yellow. There are 4 lines of dark dots on the belly scales—2 near the center and one on each outer edge. Three thin dark lines occur on the back; these are often difficult to see. Placing the snake in water may make the dorsal striping more easily visible.

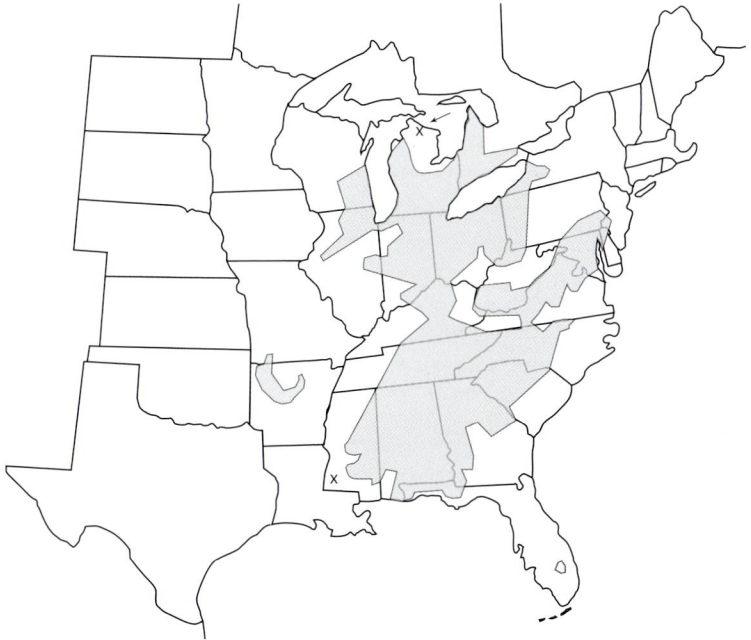

The dorsal and lateral scales are keeled and in 19 rows. The anal plate is divided.

Similar snakes: Garter snakes are usually prominently striped and have undivided anal plates. The striped crayfish snake is visibly striped; Graham's crayfish snake is slenderer and has only a single row of dark spots midventrally.

Prey: Although the primary diet of this snake consists of freshly shed (soft-shelled) crayfish, the queen snake also occasionally eats aquatic worms and insects, amphibians, and fish.

Swamp Snakes, genus *Seminatrix*

This natricine genus contains only a single species, with three subspecies. They are all of southeastern distribution. All are small and primarily aquatic snakes. As the common name indicates, these snakes are dwellers of swampy, plant-choked aquatic habitats and can be very common in some areas.

The swamp snake has smooth scales, but the scales of the first several rows above the ventral plates contain a light longitudinal line that appears superficially like a keel. The body scales are in 17 rows at midbody and the anal plate is divided.

Although these snakes may occasionally be found beneath ground surface debris, such cover is almost invariably near water.

The 3–14 live babies are virtually identical in appearance to the adults. Gravid females attain a proportionately immense girth.

Swamp snakes may wriggle energetically when captured, but are reluctant to bite. They are often easily found in the root systems of floating pond plants such as water hyacinths, water lettuce, and pennywort. During droughts, when many ponds diminish in size, swamp snakes may seek temporary to rather long-term refuge in the burrows of crayfish.

Invertebrates such as aquatic worms and leeches, as well as tadpoles, salamanders (including dwarf sirens), and small fish are the usual prey items.

129. North Florida Swamp Snake

Seminatrix pygaea pygaea

Disposition: These little natricines are harmless to humans.

Abundance/Range: Although so secretive that its presence may be unsuspected, the swamp snake can be common to abundant in suitable patches of habitat. This subspecies of swamp snake ranges northward from the approximate latitude of Tampa Bay and Vero Beach, westward through most of the panhandle to south central Alabama, and from eastern Georgia to the South Carolina state line.

129. North Florida Swamp Snake

north Florida
south Florida
Carolina

Habitat: The North Florida swamp snake occurs in cypress heads, swamps, marshes, and any other body of still or slowly moving, plant-choked water. The roots of the introduced water hyacinth are favored microhabitats. This snake may occasionally seclude itself in decaying mats of water-edge vegetation.

Numbers of these snakes move overland on warm, rainy spring and early summer nights.

Size: Most examples seen are less than 12 inches in length. A very occasional specimen may attain 18 inches. Neonates are just over 4 inches long.

Identifying features: This is a plain shiny snake that is black dorsally and brilliant orange red ventrally. Despite the overall smooth scales, the scales in the lowermost several rows have a faint, light, longitudinal, midline that, at first glance, imparts the impression that the scales are keeled. A short black bar may or may not be present at each outer edge of each red ventral plate.

Belly scale counts are the key to differentiating the races of swamp snakes. This subspecies has 118–124 belly scales while the South Florida race has fewer than 118.

The scales are arranged in 17 rows and the anal plate is divided.

Similar snakes: The red-bellied snake has a lighter dorsum and keeled scales. Red-bellied water snakes also have keeled scales and juveniles have prominent dorsal bands and a light belly.

Additional subspecies

130. The South Florida Swamp Snake, *Seminatrix pygaea cyclas*, is very similar to the above subspecies in appearance. Use range as an identification tool.

In a color echoed by numerous snakes worldwide, the South Florida swamp snake is a plain shiny black dorsally and a brilliant orange red ventrally. Despite the overall smooth scales, the scales in the lowermost several rows have a faint longitudinal midline. A short black triangle is present in both anterior corners of each red ventral plate.

130. South Florida Swamp Snake

The snake can be differentiated from the North Florida swamp snake by belly scale counts. The South Florida swamp snake has fewer than 118 belly scales while the North Florida subspecies has 118–124.

Although it is primarily a mainland species, a few have been found on barrier islands. This subspecies ranges southward to the tip of peninsular Florida from the approximate latitude of Tampa Bay and Vero Beach.

131. The Carolina Swamp Snake, *Seminatrix pygaea paludis*, ranges from the coastal plain south of Albemarle Sound, North Carolina, southward through most of eastern South Carolina to the Georgia–South Carolina state line.

The shiny black dorsum and brilliant orange-red venter of this snake are virtually identical to the colors of the two more southerly races. A short black bar is present in both anterior corners of each red ventral plate.

Whereas the two more southerly forms have 125 or fewer belly scales, the

131. Carolina Swamp Snake

Carolina swamp snake has 127 or more. Since it is difficult to impossible to count the ventral scales of a living snake, it is best to use range to identify the races.

Brown Snakes and Red-bellied Snakes, genus *Storeria*

Three species of this natricine genus occur in the United States. One species, the Florida brown snake, occurs only in Florida, but the others are very widely distributed in the eastern United States. All species in this genus have a divided anal scute, keeled dorsal and lateral scales, and all species *usually* lack loreal scales, a scale present in many snakes on each side of the head between the pre-ocular scales and the posterior nasal scale. Except for the Florida brown snake, *S. victa*, which like the red-bellied snakes has 15 scale rows, the brown snakes have 17 scale rows.

These snakes occur in open woodlands, backyards, and most habitats be-tween these extremes. They are most often found beneath debris but also bur-row readily in yielding soils.

Breeding may occur in either autumn or spring. These tiny viviparous snakes produce 3–10 (occasionally 20 or more) live young during the summer months. More northerly populations or those living at cooler elevations in the south seem to birth their young later in the year than those dwelling in warmer climes. Neonates are usually 3–4 inches long at birth.

If surprised or frightened these little snakes can appear quite feisty. They will flatten or inflate the body (thus making the light interstitial skin visible), flatten the head, and may even strike but often do so with the mouth closed. Except when on the defense this species is, except to its invertebrate prey, the picture of inoffensiveness.

Brown snakes may be surface active on relatively warm spring nights. At such times they are often seen crossing roadways.

The primary food of the northern brown snakes is earthworms, but cater-pillars, slugs, and other such fare may be opportunistically eaten. Red-bellied snakes prefer slugs.

132. Northern Brown Snake

Storeria dekayi dekayi

Disposition: This snake will often strike but almost never bites.

Abundance/Range: In bygone years, this was probably the snake most com-monly encountered by city dwellers. It seemed present under nearly every piece of discarded paper, magazine, plank, or garden stone in backyards and city parks. It was particularly common beneath debris on the sunny, grassy banks of small streams and drainage ditches. Burgeoning human populations and other urban and suburban pressures have inevitably taken their tolls on brown snake populations. This snake is found southward from eastern Maine and Ohio to the Piedmont provinces and coastal plain of North Carolina.

Habitat: Today this little snake is most apt to be encountered beneath surface debris near water, on rocky slopes, or beneath flat rocks at the edges of wood-land clearings and along fencerows. It seems absent at many of the higher eleva-tions throughout its range.

132. Northern Brown Snake

northern
marsh
Texas
Florida

Size: All subspecies of these feisty little snakes are adult at 10–14 inches in length. The record size of the northern brown snake is just over 19 inches.

Identifying features: Although the common name is quite descriptive of the dorsal coloration, the shade of brown may vary. Warm to dark brown, yellowish brown, reddish brown, or olive brown (or more rarely, gray) are all commonly seen shades. A somewhat lighter and usually well-defined vertebral stripe is often present and paralleled along each side by a row of (usually) well-separated black dots. These may occasionally be connected over the back by a thin, dark line. Dark dots and spots may or may not be present on the lower sides. The interstitial skin is whitish and may be easily visible when the snake inflates or flattens itself in fright. The ventral coloration may vary from yellowish through pale brown or whitish to pinkish. Small dark dots are usually present at the outermost extremes of the ventral scales. A dark diagonal streak is usually present behind each eye and a heavier one is usually visible on each side of the nape. Body scales are keeled, arranged in 17 rows, and the anal plate is divided.

Because they have a prominent whitish neck ring and often lack the dark spots of the adults, neonate northern brown snakes may be mistaken for a ring-necked snake.

Similar snakes: The related red-bellied snakes closely resemble the brown snakes. However, red-bellied snakes lack the diagonal dark marking posterior to the eye and have a brilliant red-orange (rarely slate gray or nearly black) venter. Smooth and rough earth snakes each have a loreal scale and pointed nose, and may have the dorsum peppered with a dusting of black dots.

Table 3. Identification features of subspecies of Brown Snakes, Storeria dekayi

Characters	dekayi	limnetes	texana	wrightorum	victa
Common name	Northern	Marsh	Texas	Midland	Florida
A short vertical streak crosses the temporal scale and extends to the lower labials	x			x	
A short dark horizontal streak restricted to the temporal scale		x			
No dark streak behind eye but a subocular spot					x
No dark streak behind the eye but dark spots on labial scales beneath the eye and at rear of jaw			x		
Adults with prominent white band at rear of head					x
Dark dorsolateral spots discrete	x	x	x		
Dark dorsolateral spots connected by dark bar				x	
15 scale rows					x
17 scale rows	x	x	x		

Comments: The various races of the brown snake can be very difficult to differentiate. Subspecific differences are based largely on facial markings and whether the dark dorsolateral dots are connected across the back by a dark line.

Examples from the periphery of a given subspecies range may have inconclusive characteristics. Despite being mentioned in Table 3, which compares the features of subspecies of brown snakes, the Florida brown snake, *Storeria victa* (account 139), is now considered a distinct species.

Additional subspecies

133. The Marsh Brown Snake, *Storeria dekayi limnetes*, is seldom found where conditions are actually wet. Rather, the snake uses suitable microhabitat on levees and dikes, at marsh edge, and in other such slightly elevated areas. It also finds areas of seclusion above the high tide mark on seashores and barrier islands.

133. Marsh Brown Snake

This race is found from the vicinity of Pensacola in Florida along the Gulf Coastal Plain to the vicinity of San Antonio Bay on Texas's Gulf Coast.

It is very much like the northern brown snake except that the marking on the temporal scale is horizontal rather than vertical. There is no dark labial barring. Neonates are very dark in overall color.

134. The Texas Brown Snake, *Storeria dekayi texana*, is found from extreme western Louisiana to central Texas and from northern Mexico to Minnesota. It lacks any marking on the temporal scale but has dark spots on the labial scales beneath the eye and at the rear of the mouth.

134. Texas Brown Snake

135. The Florida Brown Snake, *Storeria dekayi victa*, is considered a full species by some researchers. It is secretive but not uncommon. There is a possibility that imported fire ants are adversely impacting some populations of this snake. The Florida brown snake ranges over most of peninsular Florida and southeastern Georgia and a disjunct population of threatened status is found on Florida's Lower Keys. Most individuals found are 8–10 inches in length, and 12 inches seems to be the maximum length.

The scales are in 15 rows and lack apical pits. The ground color may vary from tan through various shades of brown but often has olive overtones. Reddish brown examples are not uncommon. A light vertebral stripe is usually quite evident and a row of darker spots parallels each side of this. The crown of the head is normally darker than the body color, a light collar is present, and the venter may vary from cream to a very pale pinkish white. A dark subocular spot masks the outline of the eye.

135. Florida Brown Snake

136. The Midland Brown Snake, *Storeria dekayi wrightorum*, is very much like the northern brown snake. However, this race has most of the paired dark dots on each side of the vertebral stripe connected by a dark bar over the back. This brown snake ranges southward from Michigan to Florida and from eastern Georgia to eastern Louisiana and Illinois.

136. Midland Brown Snake

137. Northern Red-bellied Snake

Storeria occipitomaculata occipitomaculata

Disposition: This snake seldom attempts to bite.

Abundance/Range: Once common beneath debris along the edges of dumps and moisture-retaining refuse heaps, red-bellied snakes now appear noticeably diminished in numbers. They seem more common at higher elevations, in areas of acidic, boggy soils, and in more densely wooded areas than the closely allied brown snakes.

The northern red-bellied snake ranges southward from the Canadian provinces of Nova Scotia, New Brunswick, southern Quebec, and Ontario's Thunder Bay region to western North Carolina, northern Georgia, and eastern Oklahoma.

Habitat: Rocks and logs both in wooded areas and at the edges of forest clearings seem especially favored. Since they do not seem to wander far, red-bellied snakes are most closely associated with areas that, while not saturated, retain the moisture necessary for the existence of their prey items.

Size: This is the smallest species of the genus. Adults of all races normally measure 8–10 inches in length; the maximum length is only 16 inches. Neonates are 3–4 inches long.

Identifying features: The various races of the red-bellied snake all have keeled scales, a divided anal plate, and scales in 15 rows.

The red-bellied snake is among the most beautifully colored of serpents. Its dorsal coloration is variable. While tan to russet seem the most commonly seen

137. Northern Red-bellied Snake

northern red-bellied
Florida red-bellied
Black Hills x northern red-bellied
Black Hills red-bellied

colors, a dark gray ground is not uncommon (at least in northern specimens), and nearly black specimens have been recorded. The belly is usually an intense orange red, but gray-bellied and black-bellied specimens have occasionally been found. No matter the basic coloration, a middorsal stripe several scales wide and often quite pale is present and usually separated from the lateral coloration by a thin darker line. This may be most prominent anteriorly. Indications of a light lateral line are sometimes present, especially on specimens with a light dorsum. Occasional dark-phase specimens bearing a broad tan to russet middorsal stripe have been reported. An (often) irregular darker stripe may occur near each side of the ventral scutes.

Characteristically, 3 light blotches are present behind the head. These include a single nape blotch and a nuchal blotch on each side. The nuchal blotches are usually narrowly separated from the ventral scales and from each other. Occasionally, however, they may touch and fuse into a light collar. A light spot is present on the upper lip below and slightly to the rear (supralabial scale 5) of each eye.

At birth red-bellied snakes are darker and more obscurely marked than the adults.

Similar snakes: The much larger size and lack of nape and nuchal spots should alone differentiate the red-bellied water snake from the diminutive red-bellied snake; juveniles of the red-bellied water snake have pale bellies and are strongly banded dorsally. Additionally, the ranges of the two do not widely overlap. The very uncommon Kirtland's snake has prominent large dorsolateral and lateral blotches and a prominent row of black dots at each side of its orange venter. The pinewoods snake has smooth scales and a white belly. Neither species of earth snake has a red-orange venter and both have a sharply pointed snout.

Additional subspecies

138. The Florida Red-bellied Snake, *Storeria occipitomaculata obscura*, is not restricted to its namesake state. It ranges northward from northern Florida to southern Arkansas and central Georgia. Rocks and logs both in wooded areas and at the edges of forest clearings, as well as backyard debris, provide cover for these tiny snakes. While tan to russet seem the most commonly seen colors, a dark gray ground may occasionally be seen. The belly is usually an intense orange red, but gray-bellied and black-bellied specimens have been found. A pale middorsal stripe *may* be present and *may* be separated from the lateral coloration by a thin darker line. If present, these markings are often most prominent anteriorly. Vague indications of a light lateral line are sometimes present, especially on specimens with a light dorsum. Characteristically, there is a light collar

138. Florida Red-bellied Snake, yellow morph

behind the dark head. The collar is formed of 3 (usually contiguous) blotches. Occasionally the blotches may fail to touch each other (this is especially true of panhandle specimens). A light spot is present on the upper lip below and slightly to the rear (supralabial scale 5) of each eye.

Although widely distributed, we have not found this snake to be common in any given area. For example, although we provide ample cover for secretive snakes in our yard, we have found only six Florida red-bellied snakes in as many years—and three of these were found in a single year. So tiny and retiring is this species that it can hide from view beneath just a few blades of grass.

139. The Black Hills Red-bellied Snake, *Storeria occipitomaculata pahasapae*, is the most poorly known, most poorly defined, and most restricted in range of the three races of red-bellied snakes. This diminutive snake has a curiously disjunct range. This may be explained by the vast expanses of unsuitable habitat that currently lie between the populations. In its pure form, the Black Hills red-bellied snake is known to occur only in central western South Dakota and adjacent Wyoming at this time. Population densities are unknown.

An intergrade population of the Black Hills x the northern red-bellied snake occurs from southern Manitoba and southeastern Saskatchewan, southward to northwestern Iowa. A disjunct population of the intergrade form exists in central southern Nebraska.

Even in its pure form, this subspecies seems poorly differentiated from the

139. Black Hills Red-bellied Snake

northern race of the red-bellied snake. It ostensibly differs by the comparative obscurity or entire lack of nape and nuchal blotches and the usual lack of a light spot on the upper lip beneath the rear of the eye. However, of the half dozen specimens we have seen from central western South Dakota, half had either relatively prominent neck blotches or blotches fused into an actual light collar. All had at least a vestige of the labial spot and two had well-developed labial spots. Both dark and light phases are known. The ventral coloration may be yellowish, orange red, or, on occasion, dark gray.

Garter Snakes and Ribbon Snakes, genus *Thamnophis*

There are nine species of this genus in eastern North America. This includes two species (eight subspecies) of ribbon snakes, and seven species (14 subspecies) of garter snakes. The defining characteristics of many species broadly overlap, sometimes rendering identification difficult. Both melanism and albinism are known. The ribbon snakes are simply slender garter snake species with very precisely delineated patterns and an affinity for aquatic habitats.

Ribbon snakes are usually reluctant to bite, but a large garter snake can prove a worthy antagonist. If grasped, all will smear musk and fecal material on their captor. As a part of the defense display, garter snakes will flatten and laterally expand their body. This not only makes them appear larger, but also displays the skin between the scales and often heightens color contrast.

Although garter snakes are considered nonvenomous snakes, the bites from some have caused adverse, though usually localized, reactions in some people bitten. All should be handled with care.

The snakes in the genus *Thamnophis* have 17–21 rows of keeled scales. The anal plate is *usually* undivided.

The garter and ribbon snakes are quite closely allied to the water snakes. In fact, some researchers have suggested that the two groups could be assigned to the same genus.

There are few habitats other than the densest woodlands and the open oceans that are not utilized by one or more species of this genus. Such habitats as rodent burrows, muskrat lodges, beneath debris, and beneath rocks and logs are favored. In optimum habitats such as trash-littered fields, dozens, even hundreds, of garter snakes may be found.

Breeding is known to occur both in the autumn and in the spring, with spring breeding (usually following the first shedding of the skin after emergence from hibernation) being most common. From as few as a half dozen to more than 90 live babies are born in mid- to late summer. The neonates measure 5–7½ inches in length and are very like the adult in appearance.

Ribbon snakes prey largely on small fish, and some garter snakes, such as the Butler's and the short-headed, eat earthworms. However, other garter snakes are opportunistic feeders, accepting all manner of amphibians, lizards, worms, leeches, slugs, and nestling mice and birds.

140. Short-headed Garter Snake

Thamnophis brachystoma

Disposition: This innocuous snake is reluctant to bite.

Abundance/Range: In suitable habitats the short-headed garter snake can be a fairly abundant species. In 1992 a researcher in Pennsylvania noted that it was not unusual to lift debris and find several dozen or more at one time. Numbers now seem diminished in many areas.

This species is restricted in distribution to western Pennsylvania and immediately adjacent New York. A disjunct population occurs in the vicinity of Tioga County, New York.

Habitat: The short-headed garter snake is a species of the meadows, fields, and hillsides of the Allegheny High Plateau. It favors areas of herbaceous growth, rock piles, and the edges of fencerows and thickets.

140. Short-headed Garter Snake

Size: This is a small, slender garter snake. Most specimens measure only 14–17 inches, and the record size is only 22 inches. Neonates are about 5¾ inches long.

Identifying features: The similarity of this small garter snake to many others may be noted by its past inclusion with the Plains, Butler's, and eastern garter snakes. As might be surmised, the identification of the short-headed garter snake can be confusing. To positively differentiate the short-headed garter snake from the Butler's garter snake you may need to count scale rows. The scales of the short-headed garter snake are in 17 rows; those of the look-alike Butler's garter snake are in 19 rows.

The short-headed garter snake has both a short nose *and* a *narrow* head. The head is barely wider than the neck. While the ground coloration is dark, it is often brownish rather than black. The 3 well-defined stripes may be whitish to a pale yellow and are usually restricted to scale rows 2 and 3 (but may, on some specimens, also involve the bottom of scale row 4). The light stripes may be highlighted by thin dark lines. Barely visible darker spots may appear in the fields of dark pigment.

Similar snakes: Both of the other northeastern garter snake species, the eastern and the Butler's, have proportionately larger heads and 19 scale rows. Ribbon snakes are more slender.

141. Butler's Garter Snake

Thamnophis butleri

Disposition: This small garter snake is not prone to bite.

Abundance/Range: The distribution of this snake in what appear to be ideal habitats seems random and curiously variable. It may be abundant at some sites but scarce or absent in others.

With a disjunct population in southeastern Wisconsin, this snake is found mainly in central, western, and northern Ohio, central and eastern Indiana, eastern Michigan, and, perhaps, adjacent extreme southern Ontario.

The identification of the Ontario specimens is questionable. Apparently the population contains a preponderance of specimens most similar externally to the short-headed garter snake, and others intermediate in appearance between Butler's and short-headed garter snakes.

Habitat: This is a species of wetlands. It inhabits stream, swamp, and marsh edges, open meadows, fields, and pastures. Butler's garter snake utilizes available surface debris—mats of grass or other vegetation, stones, paper, and boards in litter-strewn urban and suburban lots—as areas of refuge.

Size: Most Butler's garter snakes are adult at 15–20 inches in length. A few may exceed 25 inches and the record size is 27¼ inches. Neonates are about 6 inches long.

Identifying features: Although very similar to the short-headed garter snake in appearance, Butler's garter snake tends to have a richer combination of colors. The ground color may vary from an olive brown to rich black, and the light stripes may be a bright yellow or even orange. The lateral stripe includes all of scale row 3, the bottom of scale row 4, and the top of scale row 2. Some exam-

141. Butler's Garter Snake

ples have a double row of checkerboard-like alternating darker spots between the vertebral and lateral stripes. The head of Butler's garter snake is marginally wider than its neck. The keeled scales are in 19 rows, and the anal plate is undivided.

Similar snakes: The short-headed garter snake has a shorter and narrower head and only 17 rows of scales. The eastern garter snake has a proportionately broader head and the light lateral stripes do not involve scale row 4.

142. Western Black-necked Garter Snake

Thamnophis cyrtopsis cyrtopsis

Disposition: Although it may bite if carelessly restrained, this garter snake is not as apt to do so as some other species in the genus.

Abundance/Range: This is a commonly encountered snake in the Big Bend area of Texas. It is also found from southeastern Arizona northward to southeastern Colorado and southeastern Utah.

Habitat: This snake often disperses well away from permanent water sources during the rainy season. At other times it is found in rocky habitats near stock

142. Western Black-necked Garter Snake

watering tanks, rivers, streams, lakes, and marshes. It can be particularly common in rocky mountain canyons.

Size: The western black-necked garter snake typically reaches an adult length of 20–30 inches. A specimen may rarely attain 42 inches. Neonates are 8–10 inches long.

Identifying features: This subtly colored garter snake is attractive, but vastly outshone in beauty by its eastern subspecies.

The western black-necked garter snake tends to have a deep olive brown ground color between the yellow vertebral stripe and each pale yellow to whitish lateral stripe. The lateral stripes are on scale rows 2 and 3. The 3 stripes are usually brightest anteriorly. Two rows of darker checkers are usually visible, at least anteriorly. The checkers will be especially prominent if the snake is distended with food or is a heavily gravid female. Smaller black spots are often present beneath each lateral stripe. A large black blotch is present on each side of the nape. The pale-colored supralabials are prominently outlined with black pigment in the sutures. A short, stout, light, half-crescent marking is present just posterior to the rear of the mouth. This *does not* reach to the level of the top of the eye. The belly may vary from white to greenish. Babies are more brightly colored and precisely patterned than the adults.

The keeled scales are in 19 rows. The anal plate is single.

Similar species: The checkered garter snake is very similar to the western black-necked. The checkered garter snake has a tall half-crescent at the rear of

the mouth. Additionally, near the front of the snake, the light lateral line is only on scale row 3. See account 143 (next) for a description of the eastern black-necked garter snake.

Additional subspecies

143. The Eastern Black-necked Garter Snake, *Thamnophis cyrtopsis ocellatus*, a snake of the Edwards Plateau and its immediate environs, bears no truly brilliant colors, yet is remarkably pretty. The dorsal color varies from black (sometimes paler posteriorly) to olive green. There are large discrete spots anteriorly that are separated by lighter vertical bars. Encroachment on the gold to bright orange vertebral stripe and on the paler lateral stripes by the upper and lower edges of the dark blotches is normal. Posteriorly the spots may be in 2 alternating rows, but may be obscured by dark body pigment. The head is black to gray dorsally. There is a large black blotch on each side of the nape. A short, stout, light, vertically oriented marking is present just posterior to the rear of the mouth. This may be in the form of a short, stout, half-crescent, but it is often squared. There is some white on the facial scales. The white supralabial scales have broad areas of black pigment along the vertical sutures. A row of smaller black spots is present beneath each lateral line. The throat is white and the belly is white to very pale yellow green.

143. Eastern Black-necked Garter Snake

Expect this snake near the streams and springs that are found in its otherwise arid homeland.

144. Wandering Garter Snake

Thamnophis elegans vagrans

Disposition: When carelessly restrained, this snake may bite. As with any snake, it is a good idea to handle all garter snakes carefully.

Abundance/Range: This snake is generally common throughout most of its range.

This snake ranges southward from western Saskatchewan and central British Columbia to western Nevada and central New Mexico. It enters the range of this eastern reptile guide only in southwestern South Dakota and immediately adjacent Nebraska.

Habitat: The wandering garter snake occurs in most damp habitats (and some dry ones) at elevations below 10,500 feet. It may be found along the edges of streams, ponds, and marshes, in mountain and lowland meadows, in forested areas including edges and clearings, in fields, at pasture borders, and in almost all habitats intermediate between any of these. Wandering garter snakes may occasionally be found a considerable distance from water.

Size: This is an active garter snake of moderate girth and length. Adults measure 20–30 inches, with the record size being only 37 inches. Neonates are 7–8½ inches long.

144. Wandering Garter Snakes

Identifying features: The key word in any attempt at describing this snake is "variability." Almost everything about the wandering garter snake is variable, except the positioning of the lateral stripes, which are on scale rows 2 and 3. This is one of the less brightly colored garter snakes. Many specimens are actually quite dusty or dingy in appearance.

The ground color is of some shade of brown, often with a greenish tinge, or is gray. There may be a quite extensive area of darker pigment on the top of the head and another on the nape. The lateral stripes are whitish to yellowish and may be rather well defined. A double row of alternating, often rounded, separated black spots is present between the lateral and vertebral stripes. If present, the vertebral stripe is the same color as the lateral stripes, usually quite straight,

and may be well defined posteriorly. Because of the encroachment of the up-permost anterior black spots the vertebral stripe may appear zigzag and poorly defined anteriorly. The belly can be grayish or greenish and is often rather liber-ally dusted with darker pigment, especially posteriorly.

The keeled scales may be in 19–21 rows; the anal plate is usually undivided, but may occasionally be divided. There are usually 8 upper labial scales.

Similar snakes: In the geographic area covered by this book, other garter snakes within the range of this one have the lateral stripes restricted to scale row 3 (an-teriorly) or 3 and 4.

145. Checkered Garter Snake

Thamnophis marcianus marcianus

Disposition: This snake will usually bite only when seriously provoked.

Abundance/Range: This is one of the most common of semi-aridland and arid-land snakes; on some warm, moonless spring evenings dozens of checkered garter snakes may be seen crossing roadways paralleling irrigation canals. Later in the year, following parturition, even greater numbers, most of them neonates, may be found.

The range of the checkered garter snake is from southwestern Kansas to

145. Checkered Garter Snake

southeastern Arizona (a disjunct population exists along the Arizona state line in southeasternmost California) and then eastward to central Texas. This snake also occurs far southward into Mexico.

Habitat: Unlike the look-alike western black-necked garter snake that prefers rocky deserts, the checkered garter snake is often associated with sandy regions. It is still very much a creature dependent on surface water for attracting its amphibian prey.

Size: The checkered garter snake is adult at 16–28 inches. A few females may reach a length of 32 inches. The largest example yet recorded was 42½ inches long. Neonates measure 6–9½ inches in length.

Identifying features: This is one of the more variably yet subtly colored garter snakes. The ground color may be tan, pale yellow, olive yellow, or gray. There are 3 rows of olive brown to black checkers—two above the poorly defined lateral line, one below. The striping varies in color from white to yellow. All 3 stripes are of the same color. The top of the upper row of dark checkers may intrude on the edges of the vertebral stripe, making the stripe seem slightly zigzag. The belly is whitish with dark spots on the outer edges. Anteriorly, the lateral stripes are only on scale row 3 but broaden posteriorly to include scale rows 2 and 3. Immediately posterior to the mouth, a white half-crescent extends upward from the level of the lower labials to the level of the top of the eye. The whitish supralabial scales bear black sutures. A large dark blotch occurs on each side of the neck just posterior to the head.

Similar species: Some examples of the western black-necked garter snake can be very difficult to differentiate from a checkered garter snake. Both species have large black nape blotches, both have strongly patterned labial scales, and both may have a checkered body pattern. However, the light-colored crescent

that is found behind the mouth of the checkered garter snake is tall, reaching or nearly reaching the level of the top of the eye. The same crescent on the western black-necked garter snake is short and thick and seldom reaches even to the level of mideye. Additionally, anteriorly the lateral stripe of the checkered garter snake is restricted to scale row 3; that of the western black-necked garter snake is on scale rows 2 and 3.

146. Orange-striped Ribbon Snake (formerly Western Ribbon Snake)

Thamnophis proximus proximus

Disposition: This snake can seldom be induced to bite.

Abundance/Range: In the more arid areas of its range, this ribbon snake and its subspecies are difficult to find during droughts when both ground moisture and prey animals are at a premium. However, during years of normal (sufficient) moisture, these snakes are commonly seen.

The range of this ribbon snake is from western Mississippi and western Wisconsin (there is a disjunct population in eastern Illinois and western Indiana), westward to western Kansas and eastern Texas.

Habitat: As with most natricines, the habitat of the orange-striped ribbon snake is firmly tied to the availability of ground water and the availability of its amphibian prey. Look for it near stock watering tanks, in riverine habitats, along marshes and swamps, and in similar moisture providing habitats.

146. Orange-striped Ribbon Snake

| orange-striped |
| aridland |
| Gulf coast |
| red-striped |

Size: This snake is adult at 20–30 inches. The greatest authenticated length was 37⅝ inches. Neonates are 9–11 inches long.

Identifying features: The various races of this ribbon snake have a pair of spots on the rear of the parietal scales. These spots are large, colorful, and often touch each other.

The orange-striped ribbon snake is a slender creature. Its lateral stripes are on scale rows 3 and 4 and usually quite clearly defined. The lateral stripes are usually a pale to rich yellow. The vertebral stripe is bright yellow to bright orange. The belly is an unmarked off-white to pale yellow. A light (often off-white) vertical mark is present immediately anterior to the eye. The orange-striped ribbon snake usually has 8 upper labial scales (occasionally 8 on one side and 7 on the other).

The keeled scales are in 19 or 21 rows; the anal plate is not divided.

Similar species: Garter snakes are stockier and have a proportionately shorter tail. The tail of a garter snake is 25 percent or less of its total length; the tail of the ribbon snakes is greater than 25 percent of its total length. The various races of the eastern ribbon snake have 7 upper labial scales and poorly defined spots on the parietal scales. The lined snake has a row of black spots along the center of the belly. Graham's crayfish snake has a black line along the center of the belly. The queen snake has a double row of black spots along the center of the belly.

Comments: This is an alert and fast snake that will retreat into the vegetation or into the water if approached. Although capable of submerging, it is not as adept at it as the related water snakes. Ribbon snakes often simply swim along the surface to safety, disappearing into emergent or waterside vegetation as quickly as possible. (This aquatic line of escape is not without danger, however, for we once saw a Guadalupe bass rise from the depths and engulf a red-striped ribbon snake of moderate size!)

Additional subspecies

147. The Aridland Ribbon Snake, *Thamnophis proximus diabolicus*, occurs in the drier parts of southern Texas, in the Trans-Pecos region, southeastern New

147. Aridland Ribbon Snake

Mexico, and northern Mexico. Ribbon snakes of questionable lineage, but looking externally like the aridland ribbon snake, occur in northeastern New Mexico and immediately adjacent Texas. Intergrades between this race and both the western and the red-striped ribbon snakes are wide spread where ranges abut.

This race of the western ribbon snake has precisely delineated but pale yellow lateral stripes and an olive yellow to orange vertebral stripe. The ground coloration varies from olive brown to brown or olive gray, and pales somewhat posteriorly. Posteriorly, the yellow lateral lines *may* be bordered above by black horizontal dashes. There *may* be a black line along each side immediately above the outer edges of the belly plates. The unpatterned belly is a pale yellow.

Although usually smaller, this race of ribbon snake has been measured at 48 inches in length.

148. The Gulf Coast Ribbon Snake, *Thamnophis proximus orarius*, is another of the larger ribbon snakes. Although most seen are 18–30 inches in length, this race can attain a full 48 inches. The range of this species parallels the coastline from eastern Louisiana to southern Texas.

This race can be quite like the aridland ribbon snake in appearance. It often has an olive brown to dark brown dorsum. The lateral stripes are pale yellow to off-white. The vertebral stripe is olive gold to gold. The first 2 or 3 labial scales on each side of the face are tan, the remainder are pale green. The neck beneath the yellow lateral stripe is often a pale olive and shades to olive on the anterior body of the snake. Some of these ribbon snakes from southwestern Louisiana have a bluish suffusion over the entire body.

148. Gulf Coast Ribbon Snake

149. Red-striped Ribbon Snake

149. The Red-striped Ribbon Snake, *Thamnophis proximus rubrilineatus*, is, arguably, the most beautiful race of either ribbon snake species. It, too, attains a maximum length of 48 inches but is usually smaller. It occurs only in interior Texas, on and surrounding the Edwards Plateau. Like other ribbon snakes, it is associated with areas of surface water where its amphibian prey are found. This ribbon snake has a black ground color, yellow lateral stripes, and a vivid red-orange vertebral stripe. The belly is off-white to yellowish.

150. Plains Garter Snake

Thamnophis radix

Disposition: Although it would prefer to flee when surprised, a Plains garter snake will bite viciously if constrained.

Abundance/Range: The main range of this species is from northwestern Indiana westward to northeastern New Mexico and western Alberta.

This snake has diminished in overall numbers as a result of extensive habitat degradation.

Habitat: The Plains garter snake is a species of the wide-open spaces and most often associated with moisture-retaining lowlands and prairie potholes. Damp meadows, fields, pastures, city parks, and even city lots may harbor sizable populations.

150. Plains Garter Snake

Size: This garter snake attains a moderate size. Adults may measure 20–30 inches but the record size is 43 inches. It is a rather slender snake when young, but bulks up with advancing age and increasing size. Neonates are about 6¾ inches long.

Identifying features: The coloration of the Plains garter snake is less variable than that of many other species and subspecies of these striped snakes. The lateral stripes are rather high on the body (scale rows 3 and 4) and are narrow and usually paler than the well-defined orange to olive yellow vertebral stripe.

The ground color may be brownish, olive, sometimes reddish or greenish over charcoal. The black spots between the lateral and vertebral stripes may be in either a single row or an alternating double row. On the snakes with the darkest ground color, the dark spots may be nearly or entirely obscured. Another row of dark spots is present *below* the lateral line. There are prominent black vertical bars on the upper labial scales.

The belly is bluish, greenish, or yellowish with an irregular row of dark markings along each side at the anterior corner of each scute.

The scales are keeled and usually in 21 rows. The anal plate is undivided.

Similar snakes: Other garter snakes within the range of the plains garter snake have the lateral lines on scale rows 2 and 3, and many lack black bars on the upper labial scales. Ribbon snakes are very slender and attenuate and have a vertical white mark immediately anterior to the eye. The lined snake is tiny and, besides having the lateral stripes on scale rows 2 and 3, has a double row of prominent black half-moons for the full length of the venter.

151. Eastern Ribbon Snake

Thamnophis sauritus sauritus

Disposition: This harmless snake rarely, if ever, attempts to bite.

Abundance/Range: This is the most widely distributed of the various subspecies of the eastern ribbon snake. It ranges southward from southeastern Maine to Florida's panhandle and eastern Louisiana then north along the Mississippi drainage to Indiana.

In moist habitats where prey is ample, this can be a common snake.

Habitat: Look for this snake along the edges of bogs, ponds, lakes, canals, wet ditches, swamps, and marshes and at the edges of meadows and woodlands.

151. Eastern Ribbon Snake

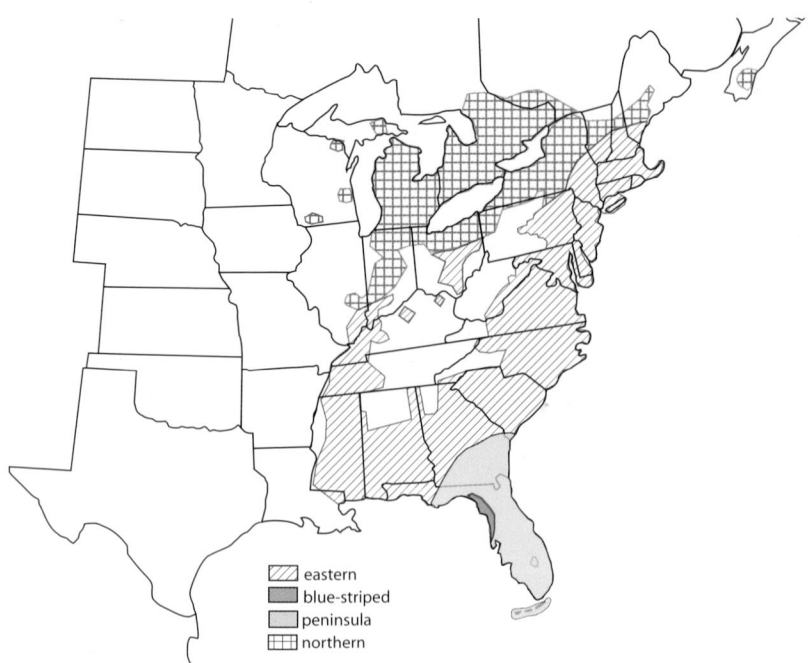

eastern
blue-striped
peninsula
northern

Size: This ribbon snake is adult at 30 inches but occasionally attains 36 inches in length. Neonates are about 8 inches long at birth.

Identifying features: This is a vividly and precisely marked ribbon snake. The ground coloration of this race is olive tan to brown. Both lateral stripes and the vertebral stripe are well defined and butter yellow to greenish yellow. Lateral stripes are on scale rows 3 and 4. The unmarked belly is often a pale yellowish green. There may be a light spot on the rear of both parietal scales, but if present it is small. A light vertical marking is present immediately in front of each large eye. There are usually 7 supralabial (upper lip) scales.

The scales are keeled, in 19 rows, and the anal plate is entire.

Similar snakes: The eastern garter snake has the lateral stripes on scale rows 2 and 3 and tends to be stockier and less precisely patterned. The blue-striped and peninsula ribbon snakes lack a well-defined vertebral stripe that contrasts strongly with the ground color. The northern ribbon snake has a ground color of dark brown or black. The various western ribbon snakes (see account 146) have well-defined parietal spots.

Comments: When moving slowly, the ribbon snakes often carry the head well above ground level, apparently in an effort to visualize their surroundings and to watch for movement by potential prey. In this, they seem more like racers than their close relatives the garter snakes do.

These snakes may be active until well after cool autumn nights and shorter days have sent most other reptiles in search of hibernacula. They also emerge from hibernation early in the spring, sometimes early enough to hunt late-chorusing spring peepers in the just greening marshlands.

Additional subspecies

152. The Blue-striped Ribbon Snake, *Thamnophis sauritus nitae*, is a rather poorly defined race. As described, it is restricted in distribution to Florida's Big Bend (Gulf Hammock) region. However, ribbon snakes with a similar amount of blue in the striping actually occur as a percentage of the population as far south on Florida's Gulf Coast as the Everglades.

The blue-stripe is a fairly small race of the eastern ribbon snake. Most examples seen are in the 20–24-inch range, and a 30-inch specimen would be considered gargantuan. Neonates are about 6 inches long.

The ground coloration of the blue-striped ribbon snake is black or a very dark brown. The lateral stripes are well defined and vary in coloration from quite a bright blue to a bluish white. The vertebral stripe is usually poorly defined, being only a shade or two lighter than the ground color for most of its length. If blue appears in the vertebral stripe it will often be only as a short streak on the nape. There may be a light spot on the rear of both parietal scales. The typical light vertical marking is present immediately in front of each eye.

The scales are keeled, in 19 rows. This small, common ribbon snake is found along the edges of marshes (both brackish and freshwater), canals, creeks, swamps, and ponds. It is often seen thermoregulating in the evening on the warmth-retaining pavement of roadways.

152. Blue-striped Ribbon Snake

153. Peninsula Ribbon Snake

153. The Peninsula Ribbon Snake, *Thamnophis sauritus sackenii*, is the southeast-ernmost race. This snake ranges northward from Florida's Lower Keys through-out the peninsula, to the eastern panhandle and to southeastern South Carolina. It is replaced in the Gulf Hammock area by the blue-striped ribbon snake.

The peninsula ribbon snake is quite common. It is adult at a length of about 30 inches, but may occasionally near 40 inches. Neonates are about 8 inches long. This, like other races of the eastern ribbon snake, is usually associated with damp areas. If startled, it glides quickly to the safety of tangles of vegetation, or may occasionally seek shelter beneath surface debris or in the water.

This is the palest and often the least precisely marked of the eastern ribbon snakes. Although the ground coloration of the peninsula ribbon snake may be dark brown, it is often considerably lighter—even tan on occasion. The white to pale yellow lateral stripes are usually reasonably well defined. The vertebral stripe can be of variable definition—easily distinguished on some individuals and all but absent on others. The unmarked belly is off-white to pale yellow. A light vertical marking is present immediately in front of each large eye.

The scales are keeled, in 19 rows.

154. The Northern Ribbon Snake, *Thamnophis sauritus septentrionalis*, is adult at 34 inches or less. Neonates are about 8 inches long. The range of this rib-bon snake extends westward from coastal southern Maine through central New Hampshire, Vermont, and most of New York to Michigan and, in disjunct pop-ulations, eastern Wisconsin. It can be common in suitable habitats.

The ground coloration of the northern ribbon snake is black or a very dark brown. Both lateral stripes and the vertebral stripe are well defined. The two lat-eral stripes are whitish yellow to butter yellow or greenish white. The vertebral stripe may be colored identically or partially suffused with darker pigment. The

154. Northern Ribbon Snake

unmarked belly is usually almost the same color as (sometimes a little greener than) the lateral stripes. The lateral stripe is separated from the ventral color by a dark stripe on scale rows 1 and 2. This dark field also includes the outer edge of each ventral scute. There may be a light spot on the rear of both parietal scales. A light vertical marking is present immediately in front of each large eye.

The scales are keeled, in 19 rows.

155. Eastern Garter Snake

Thamnophis sirtalis sirtalis

Disposition: When on the defensive, many eastern garter snakes will bite readily and repeatedly. In rare cases bites have caused localized reactions such as lividity and edema.

Abundance/Range: Although this garter snake may be difficult to find during years of drought, it remains one of the most common snakes of the eastern United States and Canada.

It ranges northward from South Florida (except on the upper Gulf Coast of the peninsula where it is replaced by the blue-striped garter snake) throughout virtually all of the United States east of eastern Texas and central Minnesota. It also occurs in southern Canada east of western Ontario.

Habitat: The eastern garter snake epitomizes the term "habitat generalist." This snake may be encountered in wet or dry prairie habitat, open mixed or hardwood woodlands, along creek and river edges, foraging in brackish and fresh water marshes, at swamp edge, along canals, and sunning on roadways.

Size: Adult at 18–30 inches, the very largest examples of this widespread race of garter snake may attain 48 inches in total length. Neonates (of all races) are about 7–9 inches long.

155a. Eastern Garter Snake, striped morph

155b. Eastern Garter Snake, striped and checkered

155c. Eastern Garter Snake, black morph

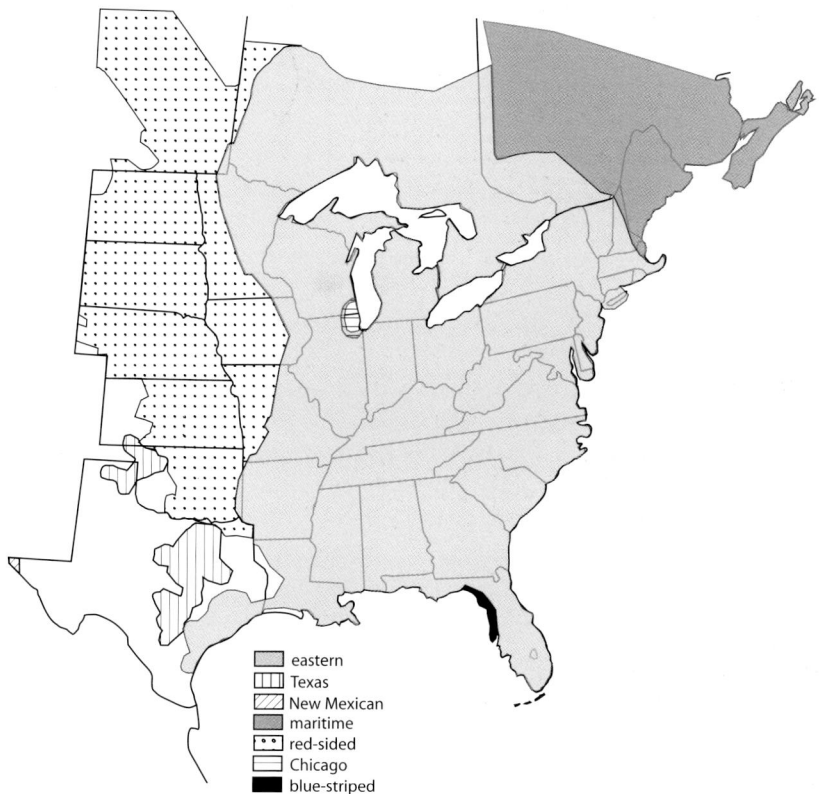

eastern
Texas
New Mexican
maritime
red-sided
Chicago
blue-striped

Identification: The ground color of this variable snake may be tan, olive, black, or orange. The pattern may consist of checkers or stripes or a combination of both, and it may be well defined or almost nonexistent. If stripes are present, the lateral stripes will be on scale rows 2 and 3. The belly is off-white to yellowish or yellow green, usually with some evidence of paired black spots on the anterior edge of each ventral scute. There are vertical black markings on the supralabial scales.

The scales are keeled, in 19 rows, the anal plate is undivided, and there are 7 upper labial scales.

Similar snakes: The ribbon snakes are slender, of "neater" appearance, and lack vertical black markings on the supralabial scales.

Comments: In one or another of its 12 races, the common garter snake (the term used for all races of *T. sirtalis* collectively) is actually represented from the Atlantic to the Pacific coasts. There are 7 subspecies in the East. It is one of North America's most widely distributed snakes. Although most races are abundant to common, a few are uncommon, and one, the San Francisco garter snake, is federally endangered.

156. Texas Garter Snake

Additional subspecies

156. The Texas Garter Snake, *Thamnophis sirtalis annectens*, occurs in two populations, one in eastern central Texas and the second in the northeastern panhandle of Texas, western Oklahoma, and extreme southwestern Kansas. This is a brilliantly marked race of the eastern garter snake.

This garter snake is found in the vicinity of permanent and ephemeral wetlands. It also utilizes virtually any other impoundment or natural surface water, and may be quite common near stock watering tanks.

The Texas garter snake has a precisely defined straw yellow to brilliant orange vertebral stripe. The lateral stripes are less well defined but very broad, involving scale row 3 as well as the halves of rows 2 and 4 adjacent to 3. The upper sides are very dark brown to black with yellow flecking. The usual size is 20–30 inches, with a record at 42¾ inches.

157. The New Mexico Garter Snake, *Thamnophis sirtalis dorsalis*, occurs only in a narrow north-south swath of land from extreme western Texas to northern New Mexico. Look for it around any existing waterholes or in riverine locations. This attractive but usually not intensely colored snake has a well-defined yellow vertebral stripe and variably defined lateral stripes. There is also a varying amount of red on the sides. On some examples the red is largely restricted to the interstitial skin (skin between the scales) but other individuals may have red on the scales themselves. The red is regularly interrupted by black spots. All colors

157. New Mexico Garter Snake. Photo by Don Sias

are best defined anteriorly and at midbody. With a record size of 51⅞ inches, this large garter snake is usually several inches shorter.

158. The Maritime Garter Snake, *Thamnophis sirtalis pallidulus*, is a variably colored snake found from northeastern Massachusetts westward through most of Quebec to the Gulf of St. Lawrence and Nova Scotia.

158. Maritime Garter Snake

It is a rather poorly understood race that reaches an adult size of 18–28 inches. The largest examples may attain a length of 3 feet.

The maritime garter snake occurs in the mature hardwood and mixed hardwood-fir forests of southeastern Canada and northern New England. Although garter snakes showing some of the characteristics of the maritime race are reported as far south as northeastern Massachusetts, the snakes from this region often show characteristics that overlap with those of the eastern garter snake.

The ground color of the maritime garter snake varies from olive gray through olive green to warm brown or almost orange. Along each side two alternating rows of black checkerboard markings stand out strongly. The vertebral line is of variable intensity. It may vary from gray to yellow (often tan) or be entirely absent, or absent posteriorly, or complete and rather well defined. The lateral lines are also of variable contrast and color, but they are usually better defined than the vertebral stripe. They are strongest along their dorsal aspect but may merge rather imperceptibly with the ventral coloration along their lower edge. The belly is light anteriorly, becoming suffused with dark pigment posteriorly.

The scales are in 19 rows.

159. The Red-sided Garter Snake, *Thamnophis sirtalis parietalis*, is a large, rather heavy-bodied subspecies. The record size for this race is 48⅞ inches. Specimens measuring 18–30 inches are considered adult. Bites from this snake have caused local redness and pain, but no overt symptoms of envenomation. As with other garter snakes, the life and habits of the red-sided garter snake are tied rather closely to the presence—at least the intermittent presence—of water. It is a snake of open woodlands, open prairies, and grasslands where it hunts its prey along creeks, ditches, potholes, marshes, swamps, and other water sources. It is

159. Red-sided Garter Snake

most commonly encountered where there is either natural or manmade ground litter under which it might hide. The range of this hardy garter snake extends southward from the southernmost regions of Canada's Northwest Territory and western Ontario to extreme northeastern Texas.

The red-sided garter snake is among the most colorful of snakes in eastern and central North America. This snake has a prominent dark-edged, yellow vertebral stripe, but the broad yellow lateral stripes are often less well defined. The lateral stripes involve all of scale rows 2 and 3 and on some specimens the bottom half of scale row 4 and the top half of scale row 1 as well. Between the lateral and vertebral lines is a series of red and black bars or a red and black checkerboard pattern. The ground color can be black, or dark to olive brown. The lateral stripe is not usually strongly separated from the olive to bluish belly.

The red-sided garter snake is primarily diurnal during cool weather and at northern latitudes, but crepuscular, or even nocturnal, during warmer weather. This is the race of garter snakes that, in some areas of Manitoba, gathers in the hundreds, even thousands, at chosen hibernacula.

160. The Chicago Garter Snake, *Thamnophis sirtalis semifasciatus*, occurs only around the southwestern tip of Lake Michigan. The subspecific name refers to the series of half bars on the anterior sides. With a record size of only 35⅞ inches, this is one of the smaller races of the common garter snake. The average adult size is 18–26 inches. This is a snake of open woodlands, city parks, undeveloped sanctuaries, and weedy urban and suburban lots. It is most abundant in areas where surface litter offers sufficient ground cover and utilizes boards, logs, woodpiles, mats of vegetation, and rocks as refugia. The Chicago garter

160. Chicago Garter Snake

snake favors wetlands such as the edges of swamps, ponds, and ditches where amphibians and earthworms are abundant.

The Chicago garter snake is very like the eastern garter snake in most characteristics. Its distinguishing characteristic is anteriorly, the lateral stripes (and, more rarely, the vertebral stripe) that are broken by 6–8 downward extending bars. The lateral stripes are on scale rows 2 and 3.

The scales are keeled, in 19 rows.

161. The Blue-striped Garter Snake, *Thamnophis sirtalis similis*, is adult at 18–28 inches. Rare individuals may near 40 inches. The blue-striped garter snake occurs along brackish marsh edges and in open woodlands along Florida's Gulf Coast from Hernando County to the Apalachicola National Forest in Wakulla County. This is a dark garter snake, with ground color varying from dark brown to black. The lateral stripes, which are well defined and on scale rows 2 and 3, are bluish white to bright blue. The yellowish vertebral stripe may be prominent to almost lacking. Vertical black markings are usually present on the rear of each supralabial scale. The belly is off-white to pale bluish white. The scales are in 19 rows.

We have seen this snake basking on early spring mornings as the warming pavement dissipated the last tendrils of fog, found it foraging in September among the emergent grasses of brackish marshes, and seen it actively crossing rain swept roadways late in October, long after darkness has fallen. The blue-striped garter snake is active throughout much of the year, throughout most of the day and far into the night.

Garter snakes with blue to bluish stripes but a *light* ground color may be

161. Blue-striped Garter Snake

found from the Everglades to Escambia County, Florida. These are a color variant of the eastern garter snake.

Lined Snake, genus *Tropidoclonion*

In overall appearance, the lined snake is much like a small garter snake; indeed, it is very closely allied to those snakes. It is distinguished from the garter snakes by the presence of a double row of bold half moons extending along the belly from the neck to the tail tip.

Lined snakes are reluctant to bite; their saliva composition is unknown. If fully provoked a lined snake may flatten its head and body and make one or more strikes at its antagonist. If defensive posturing does not deter the perceived aggressor, the lined snake will writhe once or twice and hide its head in its coils. If lifted it will smear musk and cloacal contents on the handler.

The primary and perhaps only prey of the lined snake is earthworms.

From 2 to 13 babies are produced annually.

These secretive, burrowing snakes may be found in numbers beneath sun-warmed flat rocks on barely moist, grassy hillsides.

The 19 rows of scales are keeled; the anal plate is undivided.

162. Lined Snake

Tropidoclonion lineatum

Disposition: This harmless snake rarely attempts to bite.

Abundance/Range: In spring and early summer, especially following gentle warm rains, this common to abundant snake may be surface active. It is most

162. Lined Snake

often seen in the late afternoon and well into the evening. It is so secretive that considerable populations may be present yet unsuspected.

The main continuous range of the lined snake is from central eastern South Dakota to southern Texas. Many disjunct populations exist both to the east and to the west of the main range.

Habitat: This snake is so adaptable that it may occur in backyards and vacant inner city fields, beneath roadside debris, under trash in dumps, and, of course, beneath surface rocks in grasslands and prairies.

Size: This is a moderately slender snake with a narrow head. Most examples are 7½–15 inches long. The largest yet recorded was 21½ inches. Neonates are 3½–5 inches in length.

Identifying features: Not only is this little snake a garter snake relative, but it is a garter snake look-alike as well.

Predominantly grayish to warm olive brown dorsally, the lined snake is patterned dorsally with 3 stripes. The vertebral stripe is best defined and usually the brightest yellow. The lateral stripes (one on each side on scale rows 2 and 3) are pale yellow, whitish, or merely a shade lighter than the gray or olive-based dorsal color. A row of tiny black spots may line each side of the vertebral stripe and the upper edge of each lateral line. The head is narrow and made to look even more so by the thick neck.

The belly is white, off-white, or yellowish. Two rows of prominent dark hemispheres, one on each side of the ventral midline, are present.

The keeled scales are in 19 rows and the anal plate is not divided.

Similar species: Garter snakes are the most similar but none have the double line of dark half moons along the midbelly. Graham's crayfish snake has less contrasting dorsal markings and only a single dark line on its midventer.

Earth Snakes, genus *Virginia*

This genus contains two species of sharp-nosed burrowing snakes. They prefer areas of loose soil but are most often found beneath rocks and debris in barely moist woodland, woodland edges, and woodland clearings. Of fossorial habits, snakes in sizable populations often go unnoticed.

Because preocular scales are absent, the horizontal loreal scale touches the anterior edge of the orbit. The body scales of the smooth earth snake group (the eastern, western, and mountain earth snakes) are weakly keeled, at least posteriorly (use a hand lens to check), while the scales of the rough earth snake are rather prominently keeled. Scales may be in 15 or 17 rows. The anal plate is divided. Neither species of earth snake has prominent markings, but either may be finely peppered with tiny dark dots.

A single litter of 2–8 live young is born in mid to late summer.

The preferred prey of these small snakes is earthworms. Anecdotal references to subterranean insects and slugs also being accepted as prey are numerous, but seem largely unconfirmed.

163. Rough Earth Snake

Virginia striatula

Disposition: This snake seldom bites, instead making closed-mouthed feints or hiding its head in its coils if escape attempts are foiled.

Abundance/Range: The rough earth snake has a secretive nature, making populations difficult to assess. In many regions these seem to be common snakes.

This species ranges northward and westward from north-central Florida to eastern Texas, central Missouri, and southeastern Virginia.

Habitat: The rough earth snake is a denizen of loamy pinelands and grassy verges, suburban fields, and grassy prairies. Although it often burrows, it also frequently seeks cover beneath moisture-retaining natural materials and trash. It is found with some regularity beneath boards and roofing tin on deserted homesteads when conditions are damp, but is seldom seen during times of drought.

163. Rough Earth Snake

Size: Adult at 7–9 inches in length, only rarely does this species attain 12 inches. Neonates are 2¾–4¾ inches long.

Identifying features: The dorsal and lateral color of this pencil-sized snake is usually olive gray, dark olive green, or olive brown. It may rarely be brown to almost tan and lack the olive overtones. The dorsal scales are noticeably keeled but very shiny nonetheless. The scales are in 17 rows. There are 5 upper labial scales. A horizontally oriented loreal scale is present. The venter is whitish to off-white. The neonates are often almost black in color.

Similar snakes: Eastern earth snakes have smooth or very weakly keeled scales and a more bluntly rounded snout. Brown snakes have preocular scales but lack a loreal scale. They also have a rather bluntly rounded snout. Red-bellied snakes have red, orange, or, rarely, gray or black bellies. Worm snakes have the lowermost one or two rows of lateral scales and the belly pink, and their scales are in only 13 rows.

164. Eastern Earth Snake

Virginia valeriae valeriae

Disposition: This tiny snake almost never bites.

Abundance/Range: Even where they are relatively common, eastern earth snakes are so secretive that sizable populations often go unnoticed. Because these snakes burrow deeply during drought, even diligent turning of surface debris in prime habitat during a dry period may be fruitless, but many may be found when turning the same debris following a heavy spring or summer rain. During rains these little snakes may even occasionally be encountered prowling above ground.

This species ranges northward and westward from Alachua County, Florida, to western Alabama, Ohio, and New Jersey. A disjunct colony of questionable origin exists in Highlands County, Florida.

164. Eastern Earth Snake

eastern
western
mountain

Habitat: The eastern earth snake is a denizen of grassy verges, railroad embankments, stream edges, and deciduous forest edges and openings, where it dwells beneath human-generated debris, fallen trunks, and other such naturally occurring cover.

Size: Adult at 7–9 inches, this tiny burrowing species seldom attains 12 inches in length. The record size is 15 inches. Neonates average somewhat less than 4 inches in length.

Identifying features: Dorsally and laterally the eastern earth snake may vary from gray to reddish. Its color often closely parallels the color of the soil and rocks on which it is found. The 15 rows of dorsal scales are *usually* smooth but may be very weakly keeled. There are 6 upper labial scales. The venter is whitish or may have a pale greenish sheen. The anal plate is divided. Neonates are usually much darker than the adults.

Similar snakes: Rough earth snakes have noticeably keeled scales and a very sharp nose. Brown snakes have a preocular scale but lack a loreal scale. They also have strongly keeled body scales and a rather bluntly rounded snout. Red-bellied snakes have red, orange, or, rarely, gray or black bellies. On worm snakes, the lowermost one or two rows of lateral scales and the belly are pink; their scales are in only 13 rows.

165. Western Earth Snake

Additional subspecies

165. The Western Earth Snake, *Virginia valeriae elegans*, ranges widely from western Alabama to central Texas and northward to central Iowa. It inhabits open woodlands, fields, grasslands, road verges, and the edges of open trails. Except for having 17 (rather than 15) rows of weakly keeled scales, this little snake is remarkably similar in appearance to its eastern relative.

166. The Mountain Earth Snake, *Virginia valeriae pulchra*, is every bit as tiny as the other races of this secretive snake. It seldom exceeds 11 inches in length. This race of the smooth earth snake occurs in a narrow north–south swath in

166. Mountain Earth Snake

western Pennsylvania and eastern West Virginia. The habitat of the mountain earth snake has been described as unglaciated high plateaus of the Allegheny Mountain section of the Appalachian Plateau. In this habitat the mountain earth snake is a denizen of numerous microhabitats. Rock-strewn grassy hillsides and stream edges, trash piles, forest openings, and fallen trunks are favored habitats.

Dorsally and laterally the mountain earth snake may vary from gray to reddish. It is best differentiated from the eastern earth snake by number of scale rows. The eastern earth snake has 15 scale rows for its entire body length while the mountain earth snake has 15 rows anteriorly, but 17 rows from midbody rearward. The dorsal scales are often weakly keeled. The venter is whitish or may have pale greenish sheen.

Worm, Ring-necked, Mud, and Hog-nosed Snakes, subfamily Xenodontinae

We will tentatively place these four genera in the grouping of odd-toothed snakes, the xenodontines. Some are common snakes, backyard species if you will, but others are uncommon.

About the odd-toothed ones

Some of these odd-toothed snakes are "lawn snakes." The ring-necks, for example, persist in numbers even in well-manicured, often-fertilized suburban lawns. Others, the worm snakes and hog-noses, are creatures of remote sandy fields and open woodlands. The rainbow and mud snakes are denizens of wetland habitats as diverse as cypress swamps and salt marshes.

Although most reptile enthusiasts are enamored of the little toad-hunting hog-nosed snakes with their curious turned-up noses, I (RDB) have always been enthralled by the shiny beauty and unpredictable occurrence of the mud and rainbow snakes. I remember, while searching for eastern kingsnakes along a marsh in Georgia, flipping a piece of metal lying halfway in the water and being treated to a sight of such beauty that the kingsnakes were all but forgotten. Coiled in an inch of water and half-buried in the mud lay a common rainbow snake, its vermilion, yellow, and black scales a vivid sight in my memory even after half a century.

Common rainbow snakes have never been easy for me to find, and to date the South Florida race has been an impossibility. In fact, despite concerted effort, this enigmatic snake, first found by a researcher in 1952, has never been found again.

Still, maybe someday . . .

This subfamily comprises primarily Neotropical species, all apparently with some toxic components in their saliva. Although the bite of some may cause lividity and edema at the site, no North American species is considered of medical consequence to humans. Future research may show that the familial affiliations of these snakes lie elsewhere.

Worm Snakes, genus *Carphophis*

Depending on which researcher is consulted, this genus contains either one species with three subspecies, or two full species with one (the easternmost) having two subspecies. The latter view is now rather generally accepted.

The worm snakes are fossorial. They are adept at burrowing and during droughts may follow the declining level of soil moisture far beneath the ground surface. Heavy rains may promote ground-surface activity and dispersal. On rainy springtime Massachusetts afternoons we have watched worm snakes crawl erratically from one fencerow to another, a distance of about 50 yards. During these overland excursions robins investigated but did not predate the snakes, but on one occasion a blue jay grasped and flew off with one.

Unless activated by rains, the worm snakes seem most active in the late afternoon and evening (crepuscular).

When held, worm snakes attempt escape by prodding between the fingers with both nose and tail and by smearing musk and feces on their captor. This snake does not bite when grasped.

Earthworms are the primary prey items of the worm snake. Cutworms and grubs are also eaten.

The smooth, shiny scales are arranged in 13 rows. Except when colors are muted by approaching ecdysis, a rich opalescent sheen plays over the scales. The head is small and pointed; the eyes are small. The tail terminates in a sharpened, spinous tip. The anal plate is divided. The 1–5 (rarely as many as 9) eggs are thin shelled and hatch following 7–9 weeks of incubation.

167. Eastern Worm Snake

Carphophis amoenus amoenus

Disposition: This tiny snake is entirely harmless and entirely innocuous and seldom if ever attempts to bite.

Abundance/Range: Throughout most of its range, this is a common but seldom-seen snake.

167. Eastern Worm Snake

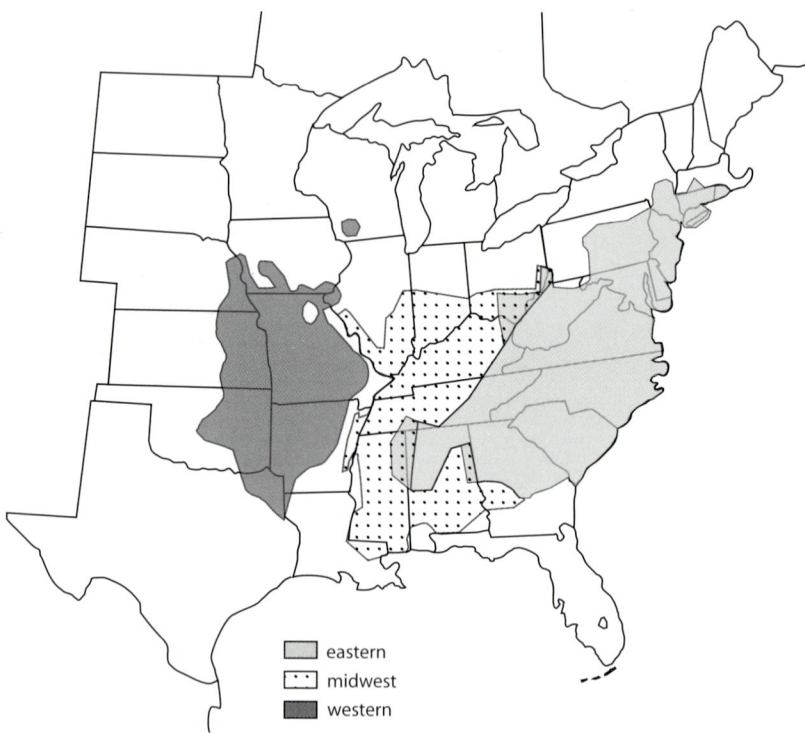

eastern
midwest
western

   The eastern worm snake ranges southward and westward from southwestern Massachusetts and southeastern New York to West Virginia, northern Georgia, and northern Alabama. It intergrades at some of the westernmost points of its range with its midwestern relative, *C. a. helenae*.

Habitat: When surface soil moisture levels are adequate, worm snakes secrete

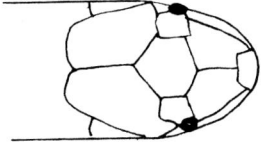 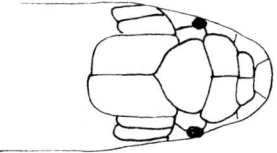

Figure 5.06. Eastern Worm Snake (left) and Midwest Worm Snake.
Illustration by Patricia Bartlett.

themselves beneath flat surface rocks or manmade debris, beneath or in rotting logs (seeming to prefer pines), or in compost piles. Dryness causes the snakes to burrow, a feat at which they are adept. This is a species of backyards, vacant lots, fields, meadows, and open woodland.

Size: With a record size of 13¼ inches, this is one of the smallest snakes of the eastern and central United States. Most specimens are considerably smaller than the record, being adult at 8–10 inches. Hatchlings are 3–4 inches long.

Identifying features: The formerly accepted common name of "pink-bellied snake" adequately describes the ventral coloration of this species. Except for the lowermost one or two rows of side scales, which are of the same pink as the venter, the shiny dorsum of the eastern worm snake is of some shade of brown (occasionally, nearly black) with a variable opalescent sheen. Snakes approaching ecdysis may be a bluish gray. The head is narrow, the eyes are tiny, and the tail terminates in a sharpened point.

The shape of certain head scales is important in differentiating the eastern from the midwestern worm snake. The eastern race has at least one of the pair (usually both) of prefrontal scales *not* fused with the internasal scales.

Similar snakes: This is the only small northeastern snake with smooth scales in 13 rows *and* with the pink belly color continuing upwards onto the first one or two rows of lateral scales. Both Kirtland's snakes and red-bellied snakes have a colorful venter of some shade of orange red (not pink or strawberry); both have keeled scales. The midwest worm snake (next account) has both prefrontal and nasal scales fused. The western worm snake is more brightly colored and the pink of its belly ascends upward to (and includes) the third row of scales.

Additional subspecies

168. The Midwest Worm Snake, *Carphophis amoenus helenae*, is found in open woodlands, on rock-strewn hillsides, and in fields, trash dumps, and similar

168. Midwest Worm Snake

habitats. It ranges southward from central Ohio and Illinois to western Georgia and eastern Louisiana. In Alabama it has been found virtually to the Florida state line along Florida's western panhandle. This shiny-scaled, warm brown and opalescent pink snake is identical in color to the eastern worm snake. The sole difference between the two is that this race has each prefrontal scale fused with the corresponding internasal scale. These scales are discrete on the eastern race.

169. The Western Worm Snake, *Carphophis amoenus vermis*, is considered a full species by some authorities. It is a common to abundant snake. It ranges eastward and southward from southeastern Iowa to extreme eastern Illinois and northern Louisiana and can be found in the moisture-preserving environs along streams. This little snake attains a length of 7–12 inches. The largest authenti-

169. Western Worm Snake

cated specimen was 14¾ inches long. The dorsum is a uniform rich brown. The belly is uniformly very bright reddish pink and the color extends upward to the third scale row above the belly scales. There is no banding or striping of any sort. The rostral (nosetip) scale is pink on the edges and underside. The smooth scales are in 13 rows.

Ring-necked Snakes, genus *Diadophis*

This genus, restricted to the United States, Canada, and northern Mexico, contains only a single species. Until recently many subspecies were generally recognized; as of 2004, however, their validities have been challenged. Certainly the characters that define subspecies are variable and overlapping; nevertheless, we have elected to retain subspecific designations here.

With only two exceptions in the United States, the members of this group are easily identified to genus by their brilliant orange to orange red neck rings. The two exceptions are the Key ring-necked snake of Florida's Big Pine Key, and the regal ring-necked snake of Texas and the West. On both, the neck ring may be faded or lacking entirely. Dependent on subspecies, the eastern members of this genus of burrowing snakes may be denizens of woodlands, plains, prairies, or even backyards.

When startled, subspecies with bright red-orange undertail (subcaudal) color coil the tail tightly and elevate it. This aposematic (warning) coloration may indicate a degree of unpalatability to predators. Predators have been known to bite, then release the snakes and rub their mouths against grasses or sand as if trying to rid themselves of an unpleasant taste. Other predators, coral snakes among them, readily eat ring-necked snakes with no sign of distress.

The 2–9 eggs are laid in the late spring and placed beneath moisture-retaining surface debris, in decomposing leaf litter, or in a burrow.

Although so secretive that they are often overlooked, ring-necks are abundant snakes. They hide beneath surface debris and even when active stay beneath recumbent grass stems and similar cover. A captured southern ring-neck will readily void feces, urates, and musk on its captor.

Salamanders, tiny frogs, equally tiny lizards, and invertebrates such as slugs (and perhaps earthworms) are eaten by this snake.

The ring-necked snakes have smooth scales and divided anal plates. The bright belly color extends upward onto the first row of body scales.

170. Southern Ring-necked Snake

Diadophis punctatus punctatus

Disposition: This common snake rarely bites. Although it does have toxic saliva that aids the snake in overcoming and predigesting prey it is entirely harmless to humans.

Abundance/Range: This common to abundant snake is found from the southernmost tip of mainland Florida northward along the coastal plain and piedmont provinces of the Atlantic states to southern New Jersey. It ranges westward to southeastern Alabama.

Habitat: Southern ring-necks are present in backyards, old fields, pastures, open woodlands, and myriad other habitats. These snakes (in fact ring-necks in general) are associated with fencerows, roadside dumps, forest openings, pond edges, and woodland edges. They dwell beneath rocks, logs, leaf mats, and other debris. These snakes may be especially common in logged areas where the stumps are left to decompose. In such situations, ring-necks may congregate behind loosened bark. Weathered tarpaper and shingle piles often provide ideal habitat for this snake.

Size: An adult size of 10–14 inches is common. The record size is just under 19 inches. Hatchlings are about 4 inches in total length.

Identifying features: The dorsal ground color of this snake is a light slate gray to mid-gray. The 15–17 rows of smooth dorsal scales have a satiny luster. The venter is yellow to orange. At the center of each ventral scute is a large black half-circle. The underside of the tail is yellow (northern snakes) to bright red

170. Southern Ring-necked Snake

southern
Key
prairie
northern
regal
Mississippi

orange (Florida peninsula snakes). Those examples with bright red-orange sub-caudal scales elevate, coil, and writhe the tail when the snake is frightened. The neck ring is well defined and usually interrupted vertebrally. The anal plate is divided. Hatchlings are darker than the adults.

Similar snakes: Brown snakes have a whitish collar, a pinkish belly, and keeled scales. Earth snakes have a pointed head, keeled scales, and a white belly. Red-bellied snakes have a light collar, keeled scales, and an unmarked red-orange belly.

Additional subspecies

171. The Key Ring-necked Snake, *Diadophis punctatus acricus*, is found beneath surface limestone flakes in clearings and on the margins of hammocks in the pine-palmetto scrubland. It is known from Big Pine Key and surrounding keys, Florida.

At an adult size of about 10 inches, this is one of the smaller races of this snake species. One tiny example, found beneath a piece of limestone on Key Largo, was only 3¼ inches in total length.

171. Key Ring-necked Snake

The smooth light slate gray to mid-gray dorsal scales have a satiny luster. The venter is yellow to orange and lacks black central spots. However, the gray of the sides encroaches on the outer edge of each ventral scute. The bright red-orange tail is coiled and writhed when the snake is frightened. The neck ring is absent or pale; if present, it is usually broken vertebrally. The color of the few hatchlings known has been paler than the color of the adults. The scales are in 15 rows.

This rare snake is protected by the state of Florida.

The tiny, introduced greenhouse frog is the primary prey item of the Key ring-necked snake, which is also known to eat slugs and reef geckos. It may feed opportunistically on other invertebrates as well.

Little is known about the reproductive biology of this snake. A 9-inch female found dead on the access road to No Name Key contained two identifiable ova. The only other reported clutch consisted of 3 eggs.

172. The Prairie Ring-necked Snake, *Diadophis punctatus arnyi,* is one of the most common snakes of our prairie states. It ranges westward from northeastern Arkansas and southeastern Wisconsin to central Texas and southeastern Colorado. The largest authenticated example was 16½ inches, but most are several inches smaller. Where the ranges abut, the prairie ring-neck readily intergrades with the regal ring-neck.

These little snakes can be found by the hundreds, literally, beneath rocks on Kansas hillsides in the spring as they emerge from their hibernacula. They seem to be almost as populous in other areas of their range. This ring-necked snake is also found beneath debris in pastures, meadows, backyards, and shortgrass

172. Prairie Ring-necked Snake

prairie. They seek the coolness of deep fissures and burrows during very hot weather.

The color of the smooth dorsal scales varies from olive gray to pale gray. The belly is yellow anteriorly, shading to orange yellow near the tail, and bright orange under the tail. There are two rows of prominent black spots on the belly. The yellow neck ring is not usually broken dorsally. This ring-neck corkscrews its tail when frightened.

This race of the ring-necked snake feeds largely on smaller snakes and lizards.

173. The Northern Ring-necked Snake, *Diadophis punctatus edwardsi*, is typically 10–14 inches long; occasional specimens may exceed 18 inches and the record size is $27^{11}/_{16}$ inches.

This is the largest race of ring-necked snake east of the Mississippi River. The range of the northern ring-necked snake extends southward from Nova Scotia and southeastern Ontario to the mountain counties of northern Alabama and Georgia.

Whether the smooth dorsal and lateral scales are slate gray, blue gray, or olive, their satiny luster imparts a look of sleekness to the ring-necks. The neck ring of the northern ring-necked snake is nearly always complete (not broken middorsally) and may be yellow, golden, or orange. The yellow to yellow orange belly often lacks all traces of black markings, but if present the markings are small, usually in a single midventral line, and missing from many scutes. The subcaudal scales (those beneath the tail) are the same color as the belly. Where

173. Northern Ring-necked Snake

intergradation with the southern ring-neck occurs in the Carolinas and Georgia, the snakes may have heavier ventral spotting, a richer orange ventral color, and a broken collar. Intergradation with the Mississippi ring-necked snake in central Alabama, western Tennessee, and western Kentucky also produces snakes with variable characteristics.

Hatchlings are darker than the adults and have a paler collar.

174. The Regal Ring-necked Snake, *Diadophis punctatus regalis*, is the largest member of this genus. The record size is 34½ inches, but most examples are considerably smaller. Where their ranges abut, the regal ring-necked snake readily intergrades with the smaller prairie ring-neck. The range of this snake extends westward from western Texas to southeastern Arizona and Utah. It also is found well into Mexico.

174a. Regal Ring-necked Snake, neck ring absent

174b. Regal Ring-necked Snake, neck ring present

The dorsum varies from slate gray to olive gray. The neck ring is often totally absent, but may be present but faded—or vividly evident. The belly is yellow anteriorly, shading to yellow orange, then red orange posteriorly. Randomly scattered black markings are present on the belly. The subcaudal scales are bright orange red. This race corkscrews the tail as a defense mechanism.

The regal ring-neck eats small snakes and lizards.

Bites from regal ring-necked snakes have caused a localized burning sensation.

Hatchlings of this large ring-neck measure about 7 inches in length.

175. The Mississippi Ring-necked Snake, *Diadophis punctatus stictogenys*, attains a length of 10–14 inches. Occasional examples may near 18 inches in length. This snake is found westward from Florida's panhandle to eastern Texas and northward to eastern Missouri.

The dorsal ground color is slate gray, blue gray, or olive. The neck ring is usually broken vertebrally. The ring may be yellow to orange. The yellow orange to orange belly usually bears a double row of small black dots. The subcaudal scales (those beneath the tail) are usually brighter orange than those of the belly. Where intergradation with the southern ring-neck occurs in eastern Alabama, the belly spotting may be more irregular. Hatchlings are darker than the adults and have a paler collar.

175. Mississippi Ring-necked Snake

This race corkscrews the tail as a defense mechanism, drawing attention from the head to the tail.

Mud Snakes and Rainbow Snakes, genus *Farancia*

The two species in this genus are specialized snakes that occur in swampy to aquatic situations. When adult, both are specialized feeders on attenuate prey items, salamanders in the case of the mud snake and eels in the case of the rainbow snake. These snakes have enlarged teeth in the rear of the upper jaw, but their salivary components are unknown.

Both species of this genus are red and black in coloration; the mud snake is banded and the rainbow snake is striped.

These snakes are of moderately heavy girth, with heavy necks and rather small heads. The tail terminates in s spinelike tip. The shiny scales are in 19 rows, the anal plate is *usually* divided. Females of both species can attain more than 5 feet in length. Males are noticeably smaller. Females either dig or enlarge existing cavities in the soil (often beneath vegetative debris) for the large clutch of eggs. An average clutch contains 25–50 eggs. However, 65–100 eggs have been reported in a single clutch for the mud snake. The females of both the mud and rainbow snakes often remain with their clutches throughout the (approximately) 60-day incubation process.

Both are mild mannered snakes that can be handled with impunity. They are aquatic and secretive, spending most daylight hours burrowed well out of sight.

At night they emerge from the mud or mulm to forage or to search for mates. Afternoon and evening showers, coupled with low barometric pressure, often induce these snakes to move overland from one water source to another.

176. Eastern Mud Snake

Farancia abacura abacura

Disposition: Although large, this snake cannot usually be induced to bite.

Abundance/Range: The eastern mud snake is one of the most secretive yet common snakes in many southeastern waterways. The eastern mud snake ranges northward from the southern tip of mainland Florida (excluding the western panhandle) to the vicinity of the Great Dismal Swamp in southeastern Virginia.

Habitat: Swamps, marshes, flooded prairies, shallow lakes (and the edges of deeper lakes), impoundments, canals, and many other aquatic habitats are populated by these aquatic serpents. They can be very common in patches of emergent and floating vegetation. Babies are often found in the root systems of water hyacinths, water lettuce, and pennywort. When not secluded in patches of aquatic plants, mud snakes are often burrowed into the substrate from which they derived their name—the mud and detritus of the pond bottom.

Size: Most mud snakes today are in the range of about 2–4½ feet. In bygone decades, it was not uncommon to find 6-foot-long females. The record size is 81½ inches. Hatchlings are about 8½ inches in length.

Identifying features: This is a robust, smooth-scaled snake with a small head and a tail tipped with a sharp spine.

The dorsal color is black. The venter (belly) is pink to red blotched liberally with black. The red coloration extends up the sides in 53 or more easily visible

176. Eastern Mud Snake

triangles. The chin is yellow to pale red spotted with black. Hatchlings are often brighter than the adults and in some cases the red may actually extend upwards across the back to form rings.

Similar species: Rainbow snakes are of similar body shape but are striped, not barred.

Comments: Although hatchlings may occasionally eat tadpoles as well as dwarf sirens, adults are specialized feeders on sirens and amphiumas and an occasional frog. Freshly captive amphiuma (voracious aquatic salamanders that may attain a yard in length) have stooled the remains of baby mud snakes, demonstrating that the tables are occasionally turned.

Additional subspecies

177. The Western Mud Snake, *Farancia abacura reinwardtii*, is slightly smaller than its eastern counterpart. Like the eastern mud snake, most adults of the western race measure 2–4½ feet. The record size, however, is 74 inches. Hatchlings are about 7 inches in length. The western mud snake is found from just east of the Florida-Alabama state line to eastern Texas and north in the Mississippi River valley to southern Illinois.

177. Western Mud Snake

The western and eastern mud snakes intergrade over much of Florida's western panhandle. The progeny may look like either of the parent races or be of roughly intermediate appearance.

It is the number of red bars and the shape of the bars that differentiate the two races of mud snakes. The red bars of the western mud snake have a squared rather than a triangular top and number 52 or fewer. The chin is yellow to pale red spotted with black.

178. Common Rainbow Snake

Farancia erytrogramma erytrogramma

Disposition: This harmless snake does not bite.
Abundance/Range: Once common, this secretive snake now seems reduced in numbers over much of its range. It is found from northern Florida (disjunct populations may still exist in Pasco and Pinellas counties) northward and westward along the coastal plains to southern Maryland and eastern Louisiana.
Habitat: The common rainbow snake is at home in spring runs, at river edge, and in shallow marshes (both fresh water and tidally influenced), bay heads (in Florida), and cypress heads.
Size: Common rainbow snakes 22–42 inches long are often seen; those measuring 48 inches or more seem less common. The record size is 66 inches. Hatchlings are about 9 inches in length.
Identification: This is a remarkably beautiful snake. The smooth scales of the back are black with 3 red lines. The lower sides are yellowish with some inter-

178. Common Rainbow Snake

common
south Florida

stitial red. The ventral scutes are red with a line of black spots along each outer edge. The tail is tipped with a sharp, spiny scale.

Similar species: Although of similar appearance, mud snakes are barred, not striped.

179. South Florida Rainbow Snake. Photo courtesy of Florida Museum of Natural History

Additional subspecies

179. The South Florida Rainbow Snake, *Farancia erytrogramma seminola*, known at the moment from only three museum specimens, is among our rarest snakes. The largest known specimen is 51½ inches in length. The average size of adults is probably somewhat less. Hatchling size is unknown.

This snake was first described five decades ago (1952) from a single specimen found in mud-bottomed Fisheating Creek, Glades County, on the west side of Florida's Lake Okeechobee. It has not been found since then. The creek is well vegetated with both emergent and floating plant species along its length. Other than these facts, nothing is known about the microhabitat of the South Florida rainbow snake.

The South Florida rainbow snake is very like the more common nominate form but darker. The dorsum is black with 3 prominent red lines. The ventro-lateral color is largely black (some yellow and red may be visible). The ventral scutes are red with a line of large black spots along each outer edge.

Note: The preserved specimen of the South Florida rainbow snake photographed for this book is in the collection of the Florida Museum of Natural History. No living specimens were available.

Hog-nosed Snakes, genus *Heterodon*

Hog-nosed snakes are among the greatest character actors of snakedom. A distressed hog-nose may attempt to hide its head in its coils—or may huff, puff, writhe, and bluff. Typical defense behaviors include spreading a "hood," flattening the head, and striking (often with the mouth closed). If the perceived aggressor doesn't leave the snake will writhe some more, open its mouth, loll it tongue out to its fullest extent, roll onto its back, and play dead. The hog-nose then shows life by immediately rolling upside down if righted.

If merely surprised but not frightened when crawling in the open, hog-nosed snakes may stop and (temperatures permitting) remain immobile for as long as they perceive danger. If grasped the snake will often writhe and smear musk and feces on its captor.

There are three species in this genus. All produce toxins in their saliva but are not aggressive toward humans.

The western hog-nosed snake eats rodents, lizards, and insects as well as amphibians. The two easternmost species are specialist feeders on toads, seldom accepting other prey.

Toads inflate themselves when frightened. Hog-nosed snakes employ their long, hinged, fanglike rear teeth to introduce toxic saliva that causes the toad to go limp and deflate, when it is then easily swallowed. In some cases the long teeth of the snake may actually puncture the body cavity and mechanically deflate the toad.

Hog-nosed snakes have enlarged adrenal glands that neutralize toxins produced by their toad prey.

The eastern forms tend to be active by day. The western forms are diurnal and crepuscular in their activity patterns. Eggs may be produced annually or biennially. Hog-noses are oviparous snakes. A clutch may consist of 4–40 eggs, but normally contains 6–20. The incubation duration is 50–68 days.

The hog-nosed snakes are stout for their length. They have a moderately to prominently upturned and dorsally keeled rostral scale, 23–25 rows of keeled body scales, and a divided anal plate.

These are snakes of open sandy areas, including sparsely wooded and agricultural sites.

180. Plains Hog-nosed Snake

Heterodon nasicus nasicus

Disposition: This snake is reluctant to bite humans, but the occasional bite has produced swelling, lividity, and some pain. Handle hog-nosed snakes carefully.

180a. Plains Hog-nosed Snake, normal morph

180b. Plains Hog-nosed Snake, reddish morph

Abundance/Range: Locally distributed and often uncommon, the plains hog-nosed snake ranges north from southern New Mexico, in a swath shaped like a reversed C, through eastern Kansas and Nebraska to central Montana and southern Alberta.

Habitat: This species prefers sandy soils in shrubby, short and mixed grass dry prairies, habitats that are becoming ever more difficult to find. Hillsides, dry sandy riverbeds interspersed with gravel, and other areas where prey is present are also suitable habitats.

Size: Although the plains hog-nosed snake has been known to attain a length of 35½ inches, the more normal size is 14–24 inches. Males are often considerably smaller than the females. Hatchlings are 6–7½ inches in length.

Identifying features: The dorsal coloration is variable. Ashy gray, tan, pale brown, or reddish may predominate. There are 5 rows of discrete, richly colored, alternating blotches. The middorsal blotches are the largest and number 35 or more on males and 40 or more on females. The first blotch on each side is elongate and angles upward to touch the edge of the parietal scale. The supralabials are light with dark pigment along their vertical sutures. Except for a white interorbital bar and a light bar crossing the head behind the eyes, the top of the head is often somewhat darker than the light body color. A broad, dark facial blotch angles downward from the eye to the rear of the mouth. There are 9 or more azygous scales (small unpaired scales) immediately behind the rostral scale. Both the belly and the subcaudal scales are predominantly black, bordered on the outer edges by white, yellow, peach, or orange. The keeled scales are in 23 rows; the anal plate is divided. See Table 4 for a comparison of identifying features of subspecies of hog-nosed snakes.

Similar species: Hook-nosed snakes are much smaller, have smooth scales, and have no keel on the top of their upturned rostral scale. The eastern hog-nosed snake has an enlarged and sharpened, but barely upturned, rostral scale.

**Table 4. Identification features of subspecies of Hog-nosed Snakes,
Heterodon nasicus**

Characters	nasicus	gloydi	kennerlyi
Common name	Plains	Dusty	Mexican
Azygous scales (scales behind the rostral scales)			
9 or more	x	x	
6 or fewer			x
Dorsal blotches on male			
32 or fewer		x	
36 or more	x		
Dorsal blotches on female			
37 or fewer		x	
40 or more	x		

Additional subspecies

181. The Dusty Hog-nosed Snake, *Heterodon nasicus gloydi*, is now considered a full species, *H. gloydi*, by some authorities. It has a narrow northeast–southwest range from south central Kansas to the Trans-Pecos region of Texas. Two disjunct populations occur in east Texas. It intergrades with the plains hog-nosed snake along the northern length of its range. Except for having fewer dorsal blotches (see Table 4), this snake is very much like the plains hog-nosed snake.

181. Dusty Hog-nosed Snake

182. Mexican Hog-nosed Snake

182. The Mexican Hog-nosed Snake, *Heterodon nasicus kennerlyi*, is found in south Texas, the Trans-Pecos region, and southwestern New Mexico and adjacent Arizona. It ranges deeply into Mexico. This race has 6 or fewer small scales (the azygous scales) posterior to the large rostral (nosetip) scale. Other than that it is similar to the dusty hog-nosed snake.

183. Eastern Hog-nosed Snake

Heterodon platirhinos

Disposition: Its slightly toxic saliva assists this snake in overcoming its toad prey. It is reluctant to bite humans, but the occasional bite has produced swell-

183. Eastern Hog-nosed Snake

183. Eastern Hog-nosed Snake

ing, lividity, and some pain. The eastern hog-nosed snake is not considered dangerous to humans.

Abundance/Range: Until the mid-twentieth century, the eastern hog-nosed snake was a common snake in most areas of suitable habitat in the eastern United States. Although the snake is still plentiful in some areas, habitat modifications and other pressures have diminished or eradicated many populations.

The eastern hog-nosed snake is found across the eastern United States from southern New England to Florida and from southeastern South Dakota to eastern Texas.

Habitat: Where there are toads, there may be eastern hog-nosed snakes. This snake is most commonly seen in areas with moderately sandy soil. Look for it in pinewoods and well-drained, open, mixed woodlands.

Size: Most examples seen are 18–30 inches long. This species occasionally attains 42 inches in length. Hatchlings are 6–7½ inches long.

Identifying features: A stocky snake, the hog-nosed snake is typified by its proportionately large head, thick neck, and short tail. The scale on the tip of the nose (the rostral scale) is large, pointed, keeled dorsally, but *not* strongly upturned. This snake occurs in a multitude of colors. The ground colors most commonly seen are light to medium brown or sandy gray. However, ground colors of red, yellowish tan, tan, grayish green, and black are also known. Black individuals begin life with typical patterns and a gray coloration. As they age, melanin suffuses the scales until the pattern is virtually overwhelmed and the entire snake is black. Examples with lighter ground colors are usually very strongly patterned, having a series of dark dorsal blotches as well as smaller alternating lateral blotches. The top of the head is dark in color. The belly is usually gray to almost black and patterned with lighter spots. The underside of the tail is distinctly lighter than the belly color. When the snake is disturbed, its tail is often rather tightly coiled. The keeled scales are in 23 or 25 rows; the anal plate is divided.

Similar species: The southern hog-nosed snake has a much more prominently upturned rostral scale than the eastern, and the underside of the southern hog-nosed snake's tail is not strikingly different than the belly color. Black racers are slender and have a satiny luster to their scales. Indigo snakes have very shiny scales. Pygmy rattlesnakes have a tiny but easily discernible tail tip rattle and facial pits. The southern copperhead has darker bands (not dorsal and lateral blotches) and facial pits.

184. Southern Hog-nosed Snake

Heterodon simus

Disposition: Like its relatives, this snake produces slightly toxic saliva. It is reluctant to bite and not dangerously venomous to humans.

Abundance/Range: This is an uncommon to rare snake.

It ranges southward and westward in the Coastal Plain from central eastern North Carolina to southern Mississippi. The range includes the northern half of the Florida peninsula. A disjunct colony is found in central Georgia.

Habitat: This small and distinctive hog-nosed snake is found in areas of sandy

184. Southern Hog-nosed Snakes

substrate. From pinewoods to meadows, and from coastal strand to turkey oak hammocks—in fact, wherever there is sandy soil—there is a chance of encountering a southern hog-nosed snake.

Size: At a record size of only 24 inches, adults of this species are much smaller than those of the eastern hog-nose. Typically, southern hog-nosed snakes are only 15–20 inches in length. Females are much the larger sex.

Identifying features: The ground of the southern hog-nosed snake may be light grayish tan, sandy gray, yellow, or terra-cotta. Strongly delineated dorsal and alternating dorsolateral blotches are present. The belly coloration is light and

may or may not be smudged with dark pigment. The underside of the tail is the same color as the belly. When the snake is disturbed, its tail is often rather tightly coiled. Like the other species in this genus, the heavy-bodied southern hog-nosed snake has a proportionately large head, a thick neck, and a short tail. The top of the southern hog-nosed snake's head is strongly patterned. The rostral scale is large, pointed, keeled dorsally, and distinctly upturned.

There are usually 25 rows of keeled scales; the anal plate is divided.

Similar species: The various races of the western hog-nosed snake, to which this species is most similar, are found far to the west. The eastern hog-nosed snake has a much less prominently upturned rostral scale than the southern, and the underside of its tail is lighter in color than its belly. Pygmy rattlesnakes have a tiny but easily discernible tail tip rattle and facial pits.

6

Venomous Snakes

Cobras, Coral Snakes, and allies; family Elapidae

Most elapid snakes are of Old World distribution, but three genera of coral snakes occur in the Americas. Only two species, eastern coral snakes and Texas coral snakes, are found within the range limitations of this book.

Coral snakes are common but secretive. They occur in habitats as diverse as isolated woodlands and suburban backyards. Despite being clad in colors that seem gaudy—rings of red, yellow, and black—these snakes are easily overlooked and blend well with their backgrounds.

Figure 6.01. Florida Cottonmouth. Illustration by K. P. Wray III.

Coral snakes have short, fixed (immovable) tubular fangs, are dangerously venomous, and should not be handled. Any bite by a coral snake should have medical attention. The venom is predominantly neurotoxic and a bite can be fatal.

Eastern Coral Snakes, genus *Micrurus*

This is the only snake genus of the family Elapidae in the eastern and central United States. It is contained in the subfamily Micrurinae. The candy cane colors are legend, although they are not unique to coral snakes. It is the *arrangement* of the colors that is usually definitive.

The coral snakes of the United States have the two warning colors of a traffic signal (red and yellow) touching each other. (This color pattern does not always hold true in Latin America, and may not hold true on the Florida Keys.) In

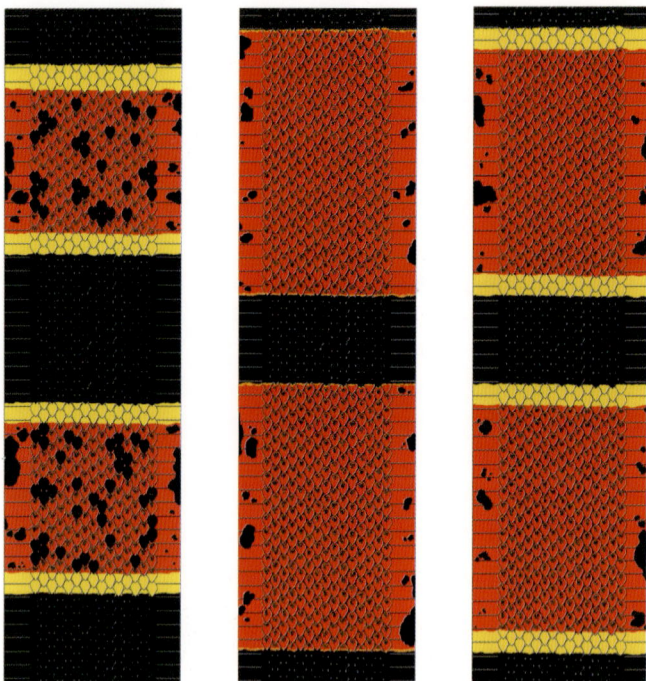

Figure 6.02. Micrurus fulvius. (a) Typical, (b) Florida Keys variant,
(c) South Florida variant. Illustrations by K. P. Wray III.

North America, except for the long-nosed snake, the nonvenomous coral snake look-alikes have the two warning colors separated by a ring of black.

There is a popular fallacy that because it has a small mouth, a coral snake must bite a finger, toe, or such to accomplish envenomation. It has also been popularly thought that a coral snake must chew to administer its venom. Holding to these two fallacies could prove fatal because neither could be further from the truth.

Because of their short rigid fangs and small mouths, coral snakes can't make the lunging strikes of the pit vipers. They don't have to aim high to bite an arm, leg, or foot, though. And although they do usually chew, a single hurried bite by the snake is all that is necessary to accomplish envenomation. Although these are comparatively small snakes, *they are not to be trifled with*. Do not handle!

This is an active, nervous, almost twitchy snake when it is above ground. When under suitable cover it is much quieter and may remain in place for several days. For example, we kept tabs on one that lived beneath a piece of plywood at the edge of our yard. It was in place every day for almost a week before it disappeared. Conversely, another lunged from its resting area when its board was lifted, disappeared into the surrounding leaves, and was never seen again.

Coral snakes prey principally on other small snakes. Lizards such as glass lizards and skinks are occasionally also eaten.

The smooth scales are in 15 rows. The anal plate is divided.

A single clutch of elongate eggs is laid annually. The clutch normally contains 3–7 eggs, but up to 13 have been recorded. The babies have a full load of venom and are capable of biting at hatching.

Despite a remarkable similarity in appearance, genetic studies have disclosed that eastern and Texas coral snakes are of different lineages and not as closely related as once thought.

Coral snakes

The coral snakes of the South Florida mainland have always been of interest to me. I (RDB) had long known that there are two color phases, the typical and the broad-banded variant, on mainland Florida. Recently I learned that a third color phase occurs on the northern Florida Keys. Of these three, only the "typical" color phase is common. It is the phase that you see along the coastal plain from southeastern North Carolina to extreme eastern Louisiana and throughout Florida.

The broad-banded variant seems quite rare and is apparently restricted in distribution to the southern tip of the Florida peninsula. Although this snake is banded in the typical combination of red, yellow, and black, its red bands are nearly twice as wide as the black bands.

My unending and as yet unfruitful search has, of course, been for this beautiful broad-banded phase. I saw this morph just once, in the possession of a Miami reptile dealer in the 1970s, and failed to photograph it. Despite a concerted search, I have not since seen another.

And what about phase three? This color morph is very different in appearance from any other coral snake in the United States and seems restricted in distribution to the northern Florida Keys. Lacking most of the yellow, it is banded in dark red and black. If yellow is present, it is usually only on the head. The presence of this red-and-black, sans most yellow, phase was brought to my attention only about five years ago.

With the finding of this color variant comes the realization that the reference often made to the colors of the coral snakes of the United States—that they are similar to those of traffic signals, with the red and yellow touching each other—is no longer valid.

Because coral snakes are very secretive, they are often thought to be rare. However, in some areas they are actually very common snakes, even dwelling in back yards.

185. Eastern Coral Snake

Micrurus fulvius

Toxicity/Disposition: Dangerously venomous. A coral snake can and will bite if restrained or molested.

Abundance/Range: This can be a common snake. Because it is so secretive, populations persist even in city parks and suburban yards, agricultural areas, and wild areas.

The eastern coral snake ranges along the Gulf and Atlantic coastal plains east of the Mississippi River, from eastern Louisiana to southeastern North Carolina. It is found throughout Florida, including Key Largo. A disjunct population exists in interior Alabama.

185a. Eastern Coral Snake

185b. Eastern Coral Snake, upper Keys variant. Photo by Jim Duquesnel

Habitat: The eastern coral snake occurs in open woodlands, hammocks, fields—even suburban backyards. It is seldom encountered only because the species is secretive, not because it is rare. This coral snake often hides under fallen tree trunks, trash, boards, mats of vegetation, or even recumbent stems of lawn grasses. It can hide effectively under seemingly sparse cover.

Size: This, the only elapid snake of the southeastern states, has a record size of 48 inches. Those most commonly seen are 20–36 inches in length. Hatchlings are about 7–9½ inches long.

Identifying features: The eastern coral snake has a somewhat flattened, relatively broad head. Its pupil is round. Its nose is usually black. Aside from aberrant colors described below, particularly on Key Largo, the *normal* sequence of ring colors on the head and neck begin with black on the snout then continue yellow-black-yellow. There is no red on the head or neck! Similarly, the tail lacks red, being ringed only in black and yellow. The rings on the body are sequenced red-yellow-black-yellow-red. The two caution colors, red and yellow, touch each other.

Many eastern coral snakes have spots of black pigment on the red scales. Although the black spotting may be random, there is often a pair of prominent black spots in each red ring.

Color aberrancies are known. Coral snakes lacking most red rings as well as snakes having inordinately wide and very bright red rings lacking black spots have been found. On Florida's Key Largo, this species often lacks most of the yellow coloration. Albinism has also been recorded.

Similar species: Use range to differentiate this snake from the Texas coral snake. No other ringed snake in the southeast has the red and yellow rings (the caution colors) touching.

186. Texas Coral Snake

Micrurus tener

Toxicity/Disposition: Dangerously venomous.

Abundance/Range: In many areas, including suburban yards and city parks, this is a common but secretive snake. It occurs west of the Mississippi River, ranging westward from western Louisiana and adjacent Missouri to the Trans-Pecos region of Texas and northeastern Mexico.

Habitat: The Texas coral snake occurs in open woodlands, hammocks, and fields. It is an accomplished burrower and spends much time underground or beneath litter.

Size: Although seldom longer than 36 inches, occasional examples may near 48 inches in length. Hatchlings are 7–9½ inches long.

Identifying features: This species is very similar to the eastern coral snake in appearance. The head is flattened and the nose is usually solid black. The sequence of ring colors beginning on the nose is black-yellow-black-yellow-red-

186. Texas Coral Snake

yellow, etc. Note that there is no red on the head or neck. Similarly, the tail lacks red, being ringed only in black and yellow. The red rings of most Texas coral snakes contain numerous scattered black spots. Color aberrancies are known.
Similar species: Use range to differentiate this snake from the eastern coral snake. No other ringed snake within the range of this species in the United States has the red and yellow rings touching each other.

Vipers; family Viperidae

This family of dangerously venomous front-fanged snakes contains two sub-families, Viperinae (true vipers) and Crotalinae (pit vipers). Only the latter subfamily, which contains the copperheads, rattlesnakes, lanceheads, and their allies, is represented in eastern and central North America.

Pit Vipers; subfamily Crotalinae

Tales have filtered down through time of monstrous rattlesnakes and immense cottonmouths. These tales have become entrenched as fact rather than fiction. The truth is, anyone suddenly confronted with the whirring rattle of a heavy-bodied rattlesnake, or by a cottonmouth swallowing a fish on a stringer, is going to think the snake is immense. A 4-foot rattler or cottonmouth can look like a 7-footer, and a 6-footer may look 10 feet long. Photographs don't always tell the truth. Taking a picture with a small-diameter lens while holding the snake (the big ones are always dead) an arm's length away will make a 5-footer look twice as big.

The greatest authenticated size recorded for a rattlesnake in the United States is an even 8 feet, for an eastern diamond-back. For the cottonmouth, the record is held by the Florida subspecies and stands at 6 feet, 2½ inches. To have seen either of these would have been a memorable occurrence for a herpetologist; by today's standards, a 4-foot rattlesnake or cottonmouth is big and a 6-footer of either is phenomenal.

The term pit viper is derived from the deep heat-sensory pits on either side of the snake's face. So sensitive are these paired facial pits that they enable a pit viper to detect the body warmth of an approaching prey animal and strike accurately in complete darkness.

The pit vipers have a long hollow fang attached to a rotatable maxillary bone on each side of the upper jaw. The maxilla can be rotated posteriorly to fold the fangs against the roof of the mouth when it is closed, or rotated anteriorly to direct the fangs almost straight forward when the snake makes a gaping, lunging, forward strike. If a fang breaks it is quickly replaced.

The fangs are ducted to venom glands at the rear of the head. It is these glands, and the controlling muscles that surround them, that form the posterior enlargement so typical of the head of these snakes.

A pit viper can regulate the amount of venom expended during a bite. Many defensive bites are "dry": no venom is injected during the strike. During other bites a full complement of venom is injected. Secondary infections can result from any bite as a result of introduced bacteria (snakes don't ever brush their teeth, and they often eat carrion).

The venom, a complex combination of proteins and enzymes, is developed primarily for prey procurement and only secondarily as a defense mechanism. The drop for drop toxicity of pit viper venom varies by species, and even by population.

All crotalines of North America give birth to live young. Studies have shown that crotalines from northern regions may produce young only every second year. Those in the South may give birth annually. The 3–20+ babies are birthed in a transparent, membranous sac from which they break free after only a few minutes.

All crotalines have elliptical pupils.

Copperheads and Cottonmouths; genus *Agkistrodon*

Both species in *Agkistrodon* should be considered dangerously venomous. In some areas of the United States the members of this genus are colloquially referred to as moccasins, with the copperheads being highland moccasins and the cottonmouths being water moccasins. Because the term moccasin has been broadened to encompass many of the water snakes, we urge that it not be used.

Actually, the terms copperhead and cottonmouth are quite descriptive. The copperheads have not only a coppery-colored to russet head, but also a similar copper banding. And you have only to see the cottony white gape of a defensive cottonmouth to understand the derivation of that name.

These snakes, both wait-and-ambush predators, accept both endothermic (warm blooded) and ectothermic (cold blooded) prey. Prey items vary from grasshoppers and cicadas to ground-dwelling birds and small mammals. A hungry copperhead may wait for prey in the same spot for days. Cottonmouths seek prey a little more actively, scavenging persistently and often being killed by vehicles as they try to eat dead animals from roadways.

These snakes have a single row of subcaudal scales and no rattles. Despite lacking a rattle, a nervous copperhead or cottonmouth can produce a very audible whirring sound by vibrating its tail in dried vegetation.

Because they often remain quietly coiled, depending entirely on their camouflage color to avoid detection, copperheads can pose a hazard when near human habitations. As might be expected, when stepped on or otherwise jostled the frightened snake will generally bite. Once riled, copperheads may strike rapidly, forcefully, and repeatedly.

Both copperheads and cottonmouths are essentially diurnal during cool weather; they adopt more nocturnal activity patterns during the hot nights of summer. They are especially active during gentle warm rains.

The young of all copperheads and cottonmouths have contrastingly colored (usually yellowish) tail tips they use as "bait," to lure in prey. The adults of some races of copperhead retain the contrasting tail color throughout their lives.

187. Southern Copperhead

Agkistrodon contortrix contortrix

Toxicity/Disposition: Dangerously venomous despite a less virulent venom than many other pit vipers. However, unless you are inordinately sensitive to the venom, a bite is not usually life-threatening. Copperheads vary in disposi-

187a. Southern Copperhead, normal morph

187b. Southern Copperhead, melanistic morph

tion. Some bite readily, others require considerable provocation before becoming defensive.

Abundance/Range: This is an abundant snake over much of its extensive range. The southern copperhead is found from southeastern North Carolina to southeastern Missouri, eastern Texas, and the northern edge of Florida's western panhandle.

Habitat: Southern copperheads occur in open woodlands, wet-bottomed ravines, woodland clearings, swamp edges, and near beaver ponds. They can be present in considerable numbers in agricultural areas and under accumulations of trash. This snake seems particularly partial to hiding beneath discarded roofing tin and pieces of plywood.

Size: Copperheads are of moderately heavy girth. Most adult southern cop-perheads are 26–34 inches in length. Occasional specimens may exceed 4 feet. Neonates are about 8–10 inches long.

Identifying features: The body scales of the southern copperhead are weakly keeled and the subcaudal scales are either entirely in a single row, or undivided anteriorly and divided posteriorly. The anal plate is undivided. The head is dis-tinctly broader than the neck. The top of the head is seldom much darker than the face. A *thin* dark line runs from the snout through the eye to the back of the head.

The ground color of the southern copperhead is tan or light brown. The snake is patterned with dark, hourglass-shaped cross bands narrowly outlined with white. The dark bands narrow dramatically dorsally (being 3 or 4 scales wide vertebrally) and are often broken and offset at midline. Laterally, the cen-ters of the dark markings are lighter than the outer edges (but darker than the ground color). Dark ventrolateral spots are present at regular intervals along the body and encroach on the outer edges of the ventral scutes. Typically, the belly is lighter than the dorsum and variably smudged with dark pigment. The facial pattern of pale tan on orange is narrowly delineated with dark pigment.

Neonates are paler than the adults and have a yellow to greenish yellow tail tip.

Similar snakes: Brightly patterned juvenile cottonmouths are often mistaken for copperheads. However, the cottonmouth has a broad dark eyestripe and darker brown bands than the southern (or other races of) copperhead. Brown water snakes have heavily keeled scales and squared dorsal blotches that are separated from the squared lateral blotches. Milk snakes have narrow heads and squared dorsal blotches that are separated from small, often rounded lateral blotches, or are ringed with black, red, and yellow. Rat snakes are blotched, not banded. Hog-nosed snakes have an enlarged, pointed, and upturned rostral (nose) scale.

Comment: Southern copperheads intergrade freely with northern copperheads along the northern periphery of their range and with the broad-banded copperhead in Texas and Oklahoma. The broad-banded copperhead intergrades not only with the southern, but with the Osage and the Trans-Pecos copperheads as well. The northern copperhead intergrades with the Osage copperhead. As might be imagined, many of the resulting snakes are often intermediate in appearance and difficult to assign to a race.

Additional subspecies

188. The Broad-banded Copperhead, *Agkistrodon contortrix laticinctus*, is a common race of central Texas and central Oklahoma. It is associated with open woodland and prairie habitats where it can be particularly common along

188. Broad-banded Copperhead

streams and near stock watering tanks. It may attain 36 inches in length but is usually several inches shorter.

The common name, of course, refers to the width of the reddish dorsal bands (which may have white edges). These are often 6 or 7 scales wide vertebrally, not dramatically narrower than on the sides. If the dark bands on the sides have lighter centers, the color difference is only moderate. The ground color is a variable tan with orangish overtones. The belly is pale tan patterned with darker mottling. The facial pattern is a light tan very precisely delineated from the darker color on the top of the head. Broad-banded copperheads of all ages have a greenish gray tail tip and the overall color of the neonates is much grayer than that of the adults.

189. The Northern Copperhead, *Agkistrodon contortrix mokasen*, occurs in varied habitats. In some areas these snakes hibernate in communal woodland dens, some at considerable elevations. Such snakes as timber rattlesnakes, black racers, and black rat snakes often den with the copperheads. After emerging from hibernation in the spring, the copperheads descend first into the talus or boulders on the slopes of their denning mountains, then disperse further into surrounding deciduous and mixed woodlands, overgrown fields, and sometimes suburban lots, as the advancing season brings additional warmth and food. This race tops out at 53 inches in length but is usually at least 12 inches shorter.

The range of the northern copperhead extends southward from southern Massachusetts and southern Illinois to the mountain counties of northern Alabama and Georgia.

189. Northern Copperhead

This race has well-defined, russet hourglass markings against an orange ground color. The bands are usually at least 3 scales wide vertebrally, darkest at the edges, and complete across the back. Rounded dark ventrolateral spots may occur in the light centers of each dark band. These may be best defined in young specimens. Dark ventrolateral spots partially cross some of the ventral scutes. Typically, the belly is lighter than the dorsum but heavily smudged with dark pigment. The top of the head is darker than the lips. A narrow dark line runs from the snout through the eye to the back of the head.

Neonates are paler than the adults and have a yellow to greenish yellow tail tip.

190. The Osage Copperhead, *Agkistrodon contortrix phaeogaster*, is a moderately sized race, with 32 inches being large and 40 inches the verified record size. This is the northwesternmost race of the copperhead. It ranges southward from extreme southeastern Iowa and extreme southeastern Nebraska to northeastern Oklahoma and southern Missouri. It is associated with rocky hillsides and fields, open woodlands, areas of second growth, streamsides, and swamp edges. It is often found beneath manmade debris, especially around deserted buildings that harbor sizable rodent populations.

The Osage copperhead is listed as Endangered by the state of Iowa, but is not protected by the other states in its range.

Although very like the northern copperhead in appearance, the Osage cop-

190a. Osage Copperhead, pallid morph

190b. Osage Copperhead, bright morph

perhead differs in normally having at least a marginally lighter ground color, lacking dark spots in the light bands, having lighter centers in the dark bands, and having the dark bands narrowly outlined in white. Adults have a greenish tail tip. The neonates are paler and have a yellow tail tip.

191. The Trans-Pecos Copperhead, *Agkistrodon contortrix pictigaster*, is, arguably, the race with the richest color. It is the southwesternmost representative of the species and occurs in the Trans-Pecos region of Texas and adjacent northern Mexico. Look for it near desert waterholes, be they rivers or stock watering tanks. It often utilizes rushes and reed at the edge of water as cover.

Having an authenticated record length of only 32⅞ inches, this is also the smallest of the copperheads.

The broad dark orange to orange-brown bands do not narrow noticeably dorsally. They are narrowly edged with white. The ground color is orange-tan. The belly is pale orange strongly patterned with mahogany bands. The outer edges of these ventral bands can be seen on the sides of the snake in the lower-center of each dark band. The tail is greenish gray. The light sides of the face are

191. Trans-Pecos Copperheads

delineated with white, separating them from the coppery hue of the rest of the head.

Neonates are less brilliantly colored than the adults.

192. Eastern Cottonmouth

Agkistrodon piscivorus piscivorus

Toxicity/Disposition: A cottonmouth of any size may signal/warn of its presence by "gaping," lying quietly coiled, head in the center of the coils but tilted back, the mouth opened very widely to show the cotton-white interior—hence the common name. The snake is quiescent at the time but it is dangerously venomous.

192. Eastern Cottonmouth

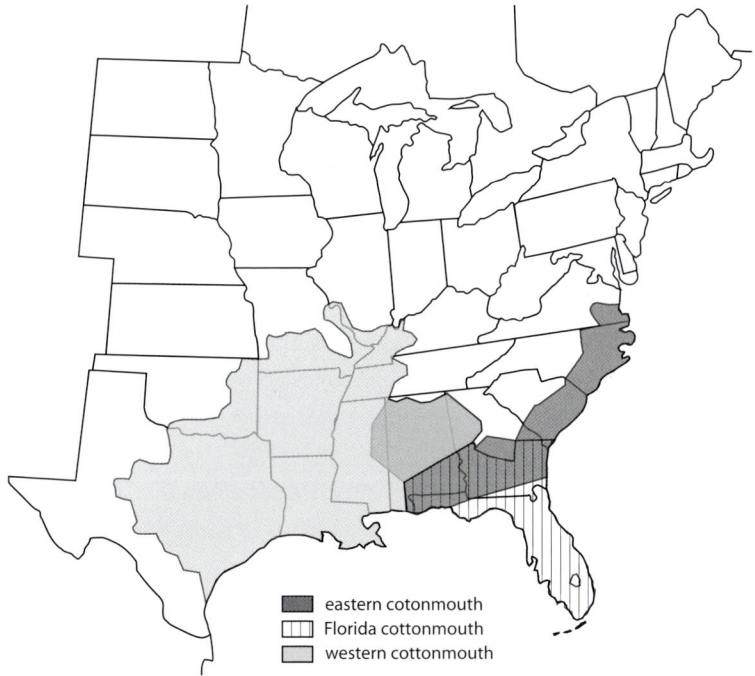

eastern cotonmouth
Florida cottonmouth
western cottonmouth

An envenomation from a big cottonmouth will be a painful, long remem-
bered experience at best, and *could* be life threatening at worst. Because the cottonmouth feeds heavily on carrion, dangerous bacteria are usually introduced
with each bite. Secondary infections may become of more concern that the
actual envenomation. A cottonmouth can deliver a venomous bite whether on
land or submerged in water. Always give this interesting snake a wide berth.

Abundance: This is a common snake throughout most of its range. It occurs on the Coastal Plain of southeastern Virginia, both Carolinas, and eastern Georgia.

Habitat: The eastern cottonmouth is almost invariably associated with water but at times may wander far from aquatic habitats. Expect it at stream- and riverside, in salt marshes, in tidally influenced estuaries, or resting on fallen tree trunks in woodland ponds. In other words, expect to see a cottonmouth whenever you are near water and sometimes when you are not.

Size: An adult size of 30–40 inches is common today. Gigantic individuals of 72+ inches have occasionally been recorded. The record length is 74 inches. Neonates average about 11½ inches in length.

Identifying features: The ground color of this heavy-bodied snake varies from olive through various browns, to almost black. Facial markings tend to be very well defined. Dark body bands are usually visible, but may be faint when the snake is dry and muddy. They are most visible when the snake is wet or just after ecdysis (skin-shedding). The body pattern will be strongest on the lower sides. The facial pattern includes a broad eyestripe that is darker than the rest of the face and bordered above and below by light pigment, tan or olive upper labials, and a pale brown or olive lower jaw marked with white.

Neonates and juveniles are very brightly colored and strongly patterned. Their eyestripe is very well defined, the top of the head is a rich brown to almost russet, and irregular edged light and medium-dark brown bands (sometimes separated by even darker bands of brown to brownish black) are present along the entire body. The tail tip is greenish yellow. Well-defined darker spots may be present in some of the body bands. The upper lip may be orangish, light brown, or olive brown.

With age and growth, a suffusion of black pigment (melanin) darkens the body (including the tail tip). The keeled scales are in 25 rows, the anal plate is entire, and the subcaudals are undivided, at least anteriorly.

Similar snakes: The southern copperhead lacks a bold facial pattern and is orange on tan. The brown, the Mississippi and Florida greens, and the banded water snakes are most often confused with the cottonmouth. All harmless water snakes have round pupils, lack facial pits, almost never stand their ground if surprised, and do not gape at a threat.

Comments: Cottonmouths can be quite bold in their actions. They may crawl slowly away when approached, may hold their ground, or even approach an antagonist, especially if the antagonist is between the snake and the water. (Harmless water snakes *may* also do this, but more often make a wide circle toward the water in an attempt to avoid the human.) When it gets close, the cottonmouth may continue into the water or it may stop and gape.

If a cottonmouth stands its ground, or actually advances toward you, use good judgment and retreat!

Whether coiled or stretched out, the cottonmouth usually holds its head tilted upward. It often vibrates its tail, which can produce a whirring sound in vegetation. Although the cottonmouth is capable of completely submerging, when not frightened a swimming cottonmouth usually holds its head well above the water and swims slowly, with its entire body on the surface. A defensive cottonmouth may retain its position, tilt its nose upward even more than normal, and gape widely, displaying the interior of the mouth.

The eastern cottonmouth intergrades freely with both the Florida cottonmouth and the western cottonmouth. Such unions may produce progeny that are impossible to assign to a subspecies.

Amphibians, other reptiles, small mammals, and small birds are all eaten, either as prey or as carrion. Vehicles kill many cottonmouths as the snakes try to eat dead animals from busy roadways.

Additional subspecies

193. The Florida Cottonmouth, *Agkistrodon piscivorus conanti*, is marginally the largest of the three cottonmouth races. An adult size of 30–36 inches is commonly seen today. Although 5-footers were not uncommon in the 1970s, they are now a rarity; large snakes are killed whenever and wherever they are found. The record size is 74½ inches. Neonates average about 11½ inches in length.

This southeasternmost race of the cottonmouth is restricted in distribution almost entirely to Florida. It occurs throughout the state except in the western panhandle, where intergrades between it and the eastern cottonmouth are found. The Florida cottonmouth is also found in extreme southeastern and extreme southwestern Georgia.

Adults of the Florida cottonmouth are heavy bodied and dark olive black to black in color. Usually the only readily visible pattern is on the face. If the snake is wet, or freshly shed, traces of a body pattern may be discernible. If present, the pattern will be strongest on the lower sides. The facial pattern involves a broad dark eyestripe, bordered above and below by light pigment, tan upper labials, and a brown and white marked lower jaw.

Neonates and juveniles are very brightly colored and strongly patterned,

193a. Florida Cottonmouth

193b. Florida Cottonmouth, juvenile

with the neonates being the brightest. Then, the eyestripe is very well defined, the top of the head is a rich brown to almost russet, and irregular edged light and medium-dark brown bands (sometimes separated by even darker bands of brown to brownish black) are present along the entire body. The tail tip is greenish yellow. Well-defined darker spots may be present in some of the body bands. The upper lip is often orangish. The keeled scales are in 25 rows, the anal plate entire, and the subcaudals undivided, at least anteriorly.

194. The Western Cottonmouth, *Agkistrodon piscivorus leucostoma*, is the smallest of the three races. It is adult at 26–40 inches and seldom exceeds 48 inches. The record size is 62 inches. Neonates are about 9½ inches in length. It is a common snake throughout most of its range, westward from extreme southeastern Mississippi to central Texas and then northward to central Oklahoma and western Kentucky. Disjunct populations exist in southern Indiana. Intergrade cottonmouths are found throughout Alabama.

This race is the darkest and has the most obscure pattern. The facial pattern may be fairly well defined but is often quite indistinct. The dorsal coloration is usually dusky black. If markings are visible they will be most prominent laterally. The belly is tan to brown smudged with black. The supraliminal (upper lip) scales are usually lighter than the face, being tan or olive. Light markings can usually be seen on the lower labial scales.

Neonates are very brightly colored and strongly patterned. Their eyestripe is very well defined, the top of the head is brown, and irregular edged light and medium-dark brown bands (sometimes separated by even darker bands of brown to brownish black) are present along the entire body. The tail tip is greenish yellow. Darker spots may be present in some of the body bands. The upper lip may be light brown or olive brown. The keeled scales are in 25 rows.

194. Western Cottonmouth

Rattlesnakes; genus *Crotalus*

The genus *Crotalus* is of New World distribution. The various species occur from the southernmost of the western Canadian Provinces in the west and central New Hampshire in the east, southward through South America to Argentina. Aruba is home to one endangered species. In the United States the genus is best represented west of the Mississippi River. East of the Mississippi there are only two species, the timber rattlesnake and the eastern diamond-backed rattlesnake.

The rattlesnakes of the genus *Crotalus* have finely fragmented crown scales. This alone will allow easy differentiation from the rattlers in the genus *Sistrurus* (the pygmy rattlers and massasaugas) that have 9 large plates on the top of their head. All have vertically elliptical pupils.

Rattlesnakes are wait-and-ambush hunters that, by chemosensory tasting with their tongue, position themselves along active rodent trails. Most strike and then immediately release the prey, allowing the venom to immobilize the stricken animal before they trail it. It has been shown that rattlesnakes are able to differentiate between the trail of a nonenvenomated animal and an envenomated one, even if of the same species.

Prairies, open woodlands and their environs, sandy scrub, and other such areas are the habitats of rattlesnakes. In rural areas these snakes may occasionally be found in vacant lots and debris-filled yards.

Rattlesnakes may not attain sexual maturity until they are several years of age. All give birth to live young. Small species may have only 2–6 babies annually (occasionally biennially) but large species such as the western diamondbacks may have 15–25 young. Both the age and the size of the adult rattlers play important parts in their reproductive biology. Females from the southernmost end of the range seem to breed annually while those from further north breed biennially, triennially, or even less frequently. Parturition occurs from midsummer to midautumn.

For the uninitiated person, an encounter with any rattlesnake is usually a memorable occurrence and if the snake is large it can be downright frightening. Although some rattlers will completely ignore the presence of a human, most (among them the western diamond-back, the Mojave, and the prairie rattlers) will react with defensive alacrity to any intrusion. Typically, the disturbed rattler will throw its body into a coil, "S" the neck and lift the head well above the coil, and rattle furiously. If you are bold enough to further near the snake, energetic strikes will be made. Beware! In most cases these snakes are earnestly attempting to dissuade additional familiarity on your part.

If undisturbed in natural habitat, you might be amazed at how little cover is needed to completely camouflage a rattlesnake of fair-size. A little dried grass, some recumbent vines and grayed twigs, and a rocky substrate will render even a large, quietly resting, rattler all but invisible. Although this remarkable camouflage stands the rattler in good stead, it can be very detrimental to the pet, livestock, or hiker that, entirely by accident, approaches too closely.

Rattlesnakes can climb (although they seldom do) and they swim well. When in the water they generally hold both head and rattle elevated above the water's surface.

These snakes are diurnally active during the cooler weather of spring or early summer, but assume a crepuscular or nocturnal pattern of activity when the hot temperatures of a desert summer set in.

Rattlers eat all manner of small mammals, but wood rats, cottontails, kangaroo rats, and pocket mice seem to figure most prominently in the diet. Desert rattlers prey readily on lizards. Ground-dwelling birds, an occasional frog, and some insects are also accepted.

Now you see them or maybe you don't

There is no feeling quite like that of strolling nonchalantly along a woodland trail, suddenly hearing the characteristic buzz of a rattlesnake—and not knowing where the sound is coming from. You stop dead in your tracks, wondering all the while how big the snake is, how upset it really is, and most importantly, whether you are within striking distance.

Happenings such as this can occur with enthusiasts and nonenthusiasts alike. The color and pattern of rattlers, even big ones, serves as an admirable camouflage when the snake is lying quietly coiled amid grasses or vines, or even in talus. This was amply and aptly demonstrated to me one autumn day as I watched a dozen or more hikers pass by a number of silent adult timber rattlesnakes, near their hibernaculum at the edge of the Appalachian Trail. The hikers never saw any of the snakes.

Clad in scales of dead-leaf brown, tan, and russet, copperheads are even more splendidly camouflaged than rattlesnakes. To fully comprehend this, you need to see (or not see) these snakes as they lie quietly in dappled shade against a background of last year's fallen leaves.

How wonderfully these snakes, wait-and-ambush predators all, are adapted for their lives of seeing but not being seen!

195. Eastern Diamond-backed Rattlesnake

Crotalus adamanteus

Toxicity/Disposition: Very dangerously venomous. An envenomation by a large example of this rattlesnake is life threatening. Some of these rattlers will coil and begin rattling at the slightest provocation while others may never make

195. Eastern Diamond-backed Rattlesnake

a sound, even if prodded. Some diamond-backs strike wildly at any nearby movement while others allow themselves to be lifted by a snake hook without any signs of bad temper.

Abundance/Range: This, the largest known rattlesnake species, ranges through-out Florida and its Keys then northward and westward to central North Carolina and eastern Louisiana.

Although it is still found generally over its entire historic range, there seem to be far fewer diamond-backed rattlesnakes today than only 10 or 20 years ago. While the loss of habitats to development and the fragmentation by roads of remaining habitats has certainly played a part in reducing the numbers of this snake, so, too, have collection for the pet trade and the skin trade, and the cyclic decline of one of the diamond-back's primary prey items—rabbits.

Habitat: This magnificent but potentially lethal snake typifies the rural South-east of bygone decades. These are snakes of piney woods and sand hills, of the Everglades and hardwood hammocks, of offshore Keys, palmetto scrublands, and coastal strands. The eastern diamond-backed rattlesnake is able to utilize a wide variety of habitats, and gives ground only grudgingly to development and its concurrent habitat fragmentation.

Size: Today, 42–54 inches is the typical size. A few decades ago 72-inch ex-amples were not uncommon; the record is an even 96 inches. Neonates average 13½ inches in length.

Identifying features: This snake is normally an attractive combination of gray on gray with black-and-white highlights. Occasionally the ground color may be olive brown. The usually cleanly defined white-outlined black dorsal diamonds (with light centers) are characteristic. So too are the 2 diagonal white lines on each side of the face. The keeled scales are in 27 or 29 rows. The tail is ringed with black and gray (or olive). Neonates are similar in pattern to the adults, but paler.

Similar species: No other southeastern snake has the dorsal diamond pattern and rattle. There are no similar species.

196. Western Diamond-backed Rattlesnake

Crotalus atrox

Toxicity/Disposition: Very dangerously venomous. This rattlesnake is capable of expelling a large quantity of potent venom, and often willing to stand its ground rather than retreat. This is a dangerous combination. The venom com-position of this species alters with the snake's growth. It is particularly adapted to cause extensive internal bleeding when the snake is a juvenile, then lessens in

196. Western Diamond-backed Rattlesnakes

virulence (but increases in volume) as the snake ages. Most western diamond-backs will coil and begin rattling at the slightest provocation. Some will strike wildly at any nearby movement. Avoid these snakes.

Abundance/Range: Despite rather concerted long-term efforts to eradicate this big rangeland rattlesnake (including collecting untold thousands during barbaric rattlesnake roundups), it remains one of the most common of snakes. It is found over most of Texas, central western Arkansas, and as far west as southeastern California. It also is widely distributed in Mexico.

Habitat: This magnificent but potentially lethal serpent typifies aridlands from Texas westward but is also found in seasonally dry habitats in Arkansas. It is

found along roadsides and in prairies, cactus and thorn-scrub deserts, canyons, pasturelands, and a wide variety of other habitats. It is occasionally found very close to city homes, and one of the few snakes that may still be encountered in vacant unmaintained suburban lots.

Size: Today, most adult western diamond-backs are 36–48 inches long. In remote areas it is still possible to find examples that measure up to 66 inches. The record size for this snake is 83⅞ inches. Neonates vary from about 8 inches in south Texas to 14 inches elsewhere.

Identifying features: These snakes are normally an interesting and attractive combination of gray on gray or gray on olive with white highlights (some from the southern Trans-Pecos are very red in color). Most are of rather pallid coloration and their pattern of only moderate contrast. The white-outlined gray or olive dorsal diamonds (with light centers) are characteristic. So too are the 2 diagonal light lines on each side of the face. The keeled scales are in 25 or 27 rows. The tail is prominently patterned with black and white (or light gray) rings of about equal width. The crown scales are strongly fragmented and the posterior light facial stripe reaches the mouth.

Neonates are similar in color to the adults but somewhat more prominently patterned.

Similar species: The Mojave rattlesnake is the only other rattlesnake that has an almost identical suite of characteristics; however, the crown scales of the Mojave rattlesnake are less fragmented than those of the western diamond-back. Additionally, on the Mojave rattler, the light stripe posterior to the eye does not touch the corner of the mouth and the black rings on the tail are narrower than the white (or gray) rings.

197. Timber Rattlesnake

Crotalus horridus horridus

Toxicity/Disposition: Dangerously venomous, but the timber rattlesnake is one of the less defensive of the rattlesnakes. Many specimens sit quietly coiled, allowing rather close approach by humans and their pets, with no trace of

197. Timber Rattlesnakes

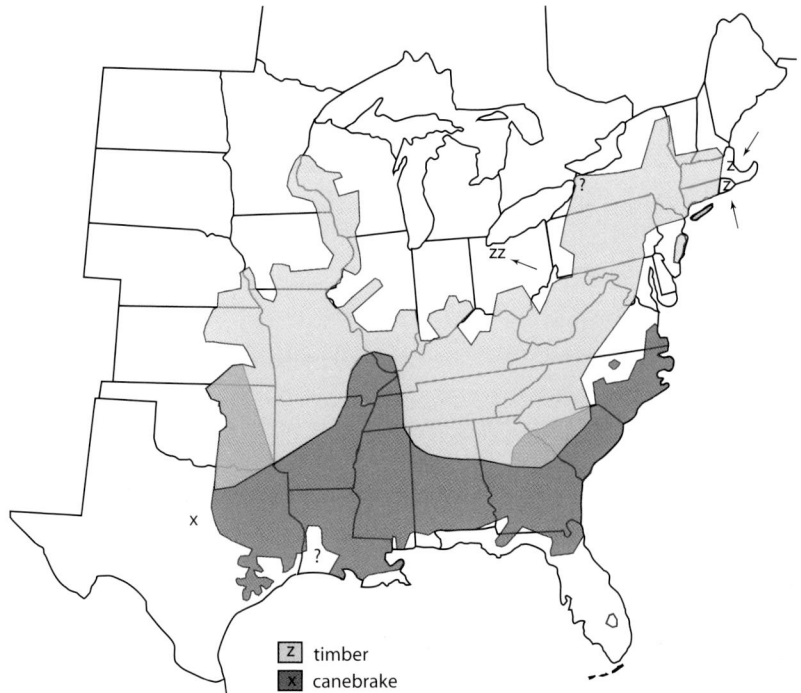

Z timber
X canebrake

hostility. Such lethargy in no way, however, negates the lethal potential of the timber rattlesnake. It is deserving of the utmost respect, and should never be unnecessarily molested.

Abundance: Because humans and rattlesnakes do not coexist well and because it is invariably the humans who win the one-sided battle for existence, timber rattlers have been extirpated from vast areas where they were once common. In fact, in many areas of the northeast, the populations of the timber rattlesnake are so diminished that protection has now been deemed necessary. However, over much of its southernmost range, this remains a relatively common rattlesnake.

The range of timber rattlesnake extends southward from central New England and northern New York, southern Ohio, and southeastern Minnesota to the Upper Piedmont Provinces of the Carolinas and northern Georgia, southern Illinois, and northeastern Texas.

Habitat: In the northeastern and central eastern states, the timber rattlesnake is typically a creature of wooded slopes, fissured mountain balds, ledges and escarpments with southerly exposures, and talus. In New Jersey, however, it is not only associated with the mountainous areas in the northern part of the state, but found in the low Pine Barrens of the south as well. Dens are usually situated in exposed elevated areas with a prevailing southern exposure. During the summer, timber rattlesnakes disperse widely from their dens, at times moving a distance of several miles.

Size: The adult size of the timber rattlesnake varies from 34 to 48 inches. In areas where food is plentiful, old adults may attain 60 inches and rarely exceed 72 inches in length. Neonates are about 11 inches long.

Identifying features: The timber rattlesnake has a ground color of black, yellow, or olive yellow. There are 15–34 broad, dark dorsal blotches or bands. Anteriorly the pattern is more typically a series of dorsal blotches that are well separated from the lateral blotches. As they near midbody the dark blotches tend to coalesce into bands.

A suffusion of dark pigment becomes visible (difficult to visualize on the black examples) about two-thirds of the way back on the snake's body, and darkens posteriorly. The rear fifth of the body and the tail may be so dark that most or all patterning is obscured. The coloration of the belly is quite like the dorsal ground color, but is variably stippled with dark pigment.

A broad dark stripe may angle downward from the eye past the back of the mouth. This may remain weakly visible throughout life or, with advancing age, obscure entirely. The facial bands tend to be least visible on yellow-phase males. A broad, often poorly defined rust to olive-colored vertebral stripe may be visible anteriorly. Neonates are similar in pattern to the adults, but paler. The ground coloration of baby timber rattlers is often a rather light olive gray or olive green. At birth it is difficult to tell which phase the snake will actually be when older.

There are 23 scale rows.

There is a broad swath of intergradation between the timber and canebrake rattlesnakes where their ranges abut.

Similar snakes: This is the only *large* rattlesnake in the northeast. The western periphery of the range of the timber rattlesnake is abutted by that of the prairie rattlesnake, and is overlapped in Oklahoma and Arkansas by the western diamond-backed rattlesnake. Both of these species have *light* stripes on the face.

Comments: By midsummer timber rattlers seem to be slowly retracing their paths to their den sites. Gravid females may give birth up to a half mile away from the den or almost exactly at the site, as we have found in western Massachusetts.

Additional subspecies

198. The Canebrake Rattlesnake, *Crotalus horridus atricaudatus,* is the southern representative of the timber rattlesnake. The venom of this race is geographi-

198. Canebrake Rattlesnake

cally variable, with that of some of the southeasternmost populations being strongly neurotoxic and particularly dangerous.

The canebrake rattlesnake has 21–29 dark bands on the body. These are usually in the form of anteriorly directed chevrons. In the southeast, the body color may vary from yellowish gray through pale brown to buff, or, rarely, a dull pink or lavender. Further westward the snakes tend to be duller and in Texas can have a ground color of medium brown (often with lighter areas outlining the dark bands) or a rather harsh olive. In all cases, the posterior third or quarter of the body becomes increasingly suffused with black pigment until the posterior one-fifth of the body and the tail are so dark that a pattern is difficult to discern. The coloration of the belly is quite like the dorsal ground color but variably stippled with dark pigment.

A broad dark line angles downward from the eye past the back of the mouth. This remains easily visible throughout life on both sexes. A broad, rust-colored vertebral stripe that is strongest anteriorly fades measurably posteriorly, but usually remains visible until obscured by the black pigment of the rear of the body.

Neonates are similar in pattern to the adults, but paler. The ground coloration of baby canebrakes is often a rather light gray. There are 25 rows of scales.

This snake is associated with forested lowlands, bramble-covered environs of lowland ponds, and swamp-edge cane stands. It does not congregate in large numbers for winter hibernation but often utilizes stump holes and other such shallow hibernacula for the dormancy. They range southward in the coastal plain and Piedmont provinces from southeastern Virginia to central Texas.

199. Mottled Rock Rattlesnake

Crotalus lepidus lepidus

Toxicity/Disposition: Dangerously venomous. The comparatively short fangs of the little mottled rock rattlesnake deliver venom that is strongly hemorrhagic in its composition. It is usually not hesitant to bite. Although probably not life threatening to a healthy human, the envenomation produces a swelling and pain that would be a decidedly unpleasant experience. If climbing among rocks or on cliff faces, carefully watch where you place your hands.

Abundance/Range: Because of this small rattler's preference for out-of-the-way habitats and the ease with which it can be overlooked, it is difficult to assess its abundance. Although it doesn't seem particularly common, one or two can usually be found by diligently searching proper habitats when climatic conditions are correct.

In the United States, the mottled rock rattlesnake is largely restricted to western Texas, from the Edwards Plateau to Hudspeth County. It also occurs in Eddy and adjacent Otero counties, New Mexico. It ranges southward to the vicinity of San Luis Potosí, Mexico.

Habitat: A snake of rocky mountainsides, cliff faces, escarpments, outcroppings,

199. Mottled Rock Rattlesnake

199. Mottled Rock Rattlesnake

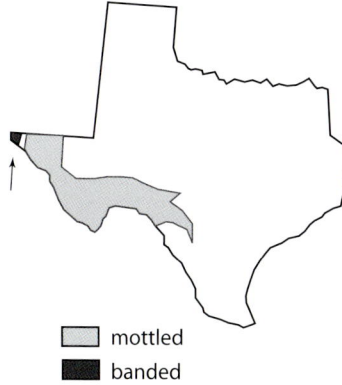

mottled

banded

and fissured canyon walls, the mottled rock rattlesnake occasionally wanders quite far afield and may be encountered considerable distances from typical habitat. It is also a well-known denizen of the many road cuts in Texas.

Size: The normal range of this small rattlesnake is 12–18 inches, with a record of 30½ inches. Neonates are about 6–8½ inches in total length.

Identifying features: Despite coming from a relatively small range, the mottled rock rattlesnake is quite variable in color. The ground color typically blends well with the color of the rocks among which the snake is found. The ground color can vary from pinkish to russet to gray to bluish gray to chalky white, and all colors in between. The primary bands are usually at least faintly visible (sometimes vividly so), but the secondary mottling is noticeably less contrasting. This snake is capable of quite considerable day-to-night color changes. The top of the head is weakly mottled or plain. A diagonal dark bar is present from beneath each eye to the corner of the mouth. The weakly banded tail is a rather bright yellow to orange at birth, but darkens somewhat as the snake ages. The keeled body scales are arranged in 23 rows. The anal plate is not divided.

Similar species: The related banded rock rattlesnake lacks facial banding. Some gray-banded kingsnakes can be very similar to the mottled rock rattler in color and pattern; however, the kingsnake has round pupils, no facial pits, and no rattle. The Texas lyre snake can be of similar color and also has vertically elliptical pupils. However this is a slender snake with neither facial pits nor rattle.

200. Banded Rock Rattlesnake

Additional subspecies

200. The Banded Rock Rattlesnake, *Crotalus lepidus klauberi*, is the more westerly representative of this species. It has been measured at a full 32 inches in length, but most examples are in the 12–18-inch range. In our coverage area, this race occurs only in the Franklin Mountains of west Texas. From there it ranges westward to southeastern Arizona and southward to Jalisco, Mexico.

The banded rock rattlesnake has a cleaner, less busy pattern than the mottled rock rattlesnake. In fact, many specimens of the banded rock rattlesnake lack even vestiges of secondary markings between the primary bands. There is a tendency toward sexual dimorphism, with many males having a ground color of green (moss green to bluish green) while that of the females is grayish to bluish. The bands are deep brown to black. This race either has no facial markings, or they are faded. The tail tip of hatchlings is bright yellow to yellow orange. This fades, but is often not entirely lost as the snakes mature.

201. Northern Black-tailed Rattlesnake

Crotalus molossus molossus

Toxicity/Disposition: Dangerously venomous. This large rattlesnake has dangerously toxic venom but usually a relatively benign disposition. Despite this perceived reluctance to bite, an envenomation by a black-tailed rattlesnake can have very serious consequences. Approach this snake with caution.

Abundance/Range: Although a species with a wide distribution, the northern

201. Northern Black-tailed Rattlesnake

black-tailed rattlesnake is not particularly common in Texas.

The range of the black-tailed rattlesnake extends westward from Edwards Plateau of Texas to western Arizona, and from southern Sonora to northern Arizona and northwestern New Mexico. There are Mexican subspecies.

Habitat: This is a species of sparsely wooded rocky canyons and mountains, talus and lava beds, desert arroyos, rocky plains, and floodplains. It has been found in arid dune lands, in boulder-strewn flatlands, in creosote–prickly pear desert lands and in coniferous and mixed woodlands. It may be considered a habitat generalist.

Size: This is a fairly large, moderately slender rattlesnake. An occasional specimen may attain just over 54 inches but most of those seen are 18–42 inches long. Neonates are 9½–12 inches in length.

Identifying features: West of our coverage area, the northern black-tailed rattlesnake is of much brighter color.

In Texas, the ground color varies from greenish brown to silvery green. Black blotches alternate with silver, silver gray, or olive silver on the dorsum and become less contrasting and more poorly defined, posteriorly. The region of the neck and anterior body bears a broad black dorsal stripe containing 6 or more pale greenish blotches. The black dorsal blotches, variable and uneven in outline, usually contain light centers (which may break the dark blotch laterally), or at least a few light scales. These blotches narrow laterally and often continue to the venter as a thin irregular dark bar. The snout, from above the eyes to the nosetip, is suffused with dark pigment. A broad diagonal dark bar extends from the crown through the eye to the angle of the mouth. It is bordered anteriorly by a broad, variably defined light bar, and may or may not be bordered posteriorly by a poorly defined light bar. The tail is black.

Neonates are quite similar but have weakly banded tails.

Similar snakes: None in our area. The dark nose, black tail, and light-spotted dark blotches present a diagnostic suite of characteristics.

202. Mojave Rattlesnake

Crotalus scutulatus scutulatus

Toxicity/Disposition: Dangerously venomous. The venom of this common rattlesnake varies geographically. In some populations the venom is predominantly neurotoxic (type A). The identifiable neurotoxin is now called Mojave-toxin. Other Mojave rattlers have a basically hemotoxic venom (type B). Those from the Big Bend region possess the more dangerous type-A venom. A bite from any Mojave rattler is very dangerous and can be life threatening. This snake will bite readily; use extreme care when anywhere near one. If bitten, seek medical help immediately.

202. Mojave Rattlesnake

Abundance: The Mojave rattlesnake is a relatively common species. Its true population status is confounded by the fact that it is often mistaken for the even more abundant western diamond-back. It enters our area only west of Big Bend in Texas. Outside of Texas it ranges westward to southern California and southward well into Mexico.

Habitat: This rattlesnake is essentially a species of rock-strewn grasslands, desert scrub areas, and rocky, sparsely vegetated mountain slopes. Plant communities with which it is associated include mesquite, ocotillo, cholla, and creosote bush.

Size: The Mojave rattlesnake is not a particularly large species, nor is it quite as heavy bodied as some other rattlers. Most are in the ranges of 18–40 inches; the largest recorded specimen measured only 51 inches. Neonates are about 10½ inches long.

Identifying features: Because of the irregular, very well-defined, dark diamond- or rhomboid-shaped markings on its back, the Mojave rattlesnake was long called the Mojave diamondback. The dorsal markings have light centers and are usually rather broadly margined with dark brown pigment. The ground color may vary from greenish yellow through tan to brownish or olive gray. Two rows of lateral blotches, the uppermost the weakest, may be present. There are 2 light, diagonal facial stripes. The aft stripe passes downward *behind* the angle of the jaws. The tail is ringed, with the dark rings irregular, often offset vertebrally, and less than half as wide as the white to light gray ones.

Typically, the Mojave rattler has a characteristic crown scalation. Rather than the scales being fully fragmented, the crown *usually* has 2 or 3 rather large scales between the supraoculars. These are followed by numerous crown scales that are proportionately larger than those on the heads of all other typical rattlesnakes. The scales are in 25 rows and the anal plate is undivided.

Similar snakes: The western diamond-backed rattlesnake has well-defined black-and-white tail rings of equal width. The black-tailed rattlesnake has both a black tail and a very dark mask. Rock rattlesnakes have either a dark facial bar or none. Admittedly, the size of crown scales is almost impossible to ascertain on a living snake, but the presence of relatively large crown scales, compared to those of other rattlers, is diagnostic for the Mojave rattlesnake.

203. Prairie Rattlesnake

Crotalus viridis viridis

Toxicity/Disposition: This is a dangerously venomous snake. Venom composition may vary across the range. This snake is irascible and usually ready to bite at the slightest provocation. Use extreme caution when near a prairie rattler.

Abundance/Range: Despite having been subjected to decades of intense persecution, the prairie rattlesnake remains common through much of its extensive range.

The easternmost representative of the species, it ranges southward from southeastern Alberta and adjacent Saskatchewan to southern Texas and northern Mexico.

Habitat: This snake is a grassland-prairie species. It hibernates in rocky outcroppings, escarpments, mountainside dens (all preferably with a southern exposure), and abandoned mammal burrows.

Size: The usual size is 18–40 inches, with the largest authenticated length at 57 inches. Neonates are 8–11½ inches in length.

Identifying features: The specific name *viridis* refers to the greenish ground color of some examples. However, the ground color may also be brownish or grayish. The dorsal blotches are usually well defined and on the anterior two-thirds of the body, edged anteriorly and posteriorly with darker pigment, and completely margined with light pigment. Posterior blotches are less well de-

203. Prairie Rattlesnake. Photo by Regis Opferman

fined. The 2 light facial stripes are usually well defined. The more posterior stripe angles sharply rearward above the jaw line and does not touch the mouth. A light bar often crosses the dorsal surface of the head from supraocular scale to supraocular scale. The tail is strongly but narrowly barred (not ringed) in olive or gray and black. The most posterior bar and basal segment of rattle are black. Neonates are more precisely patterned than adults.

The keeled scales are usually in 25 (23–27) rows and the anal plate is undivided.

Similar snakes: Both the western diamond-back and the Mojave rattler have black-and-white ringed tails. The black-tailed rattlesnake has both a black tail and a very dark mask. Rock rattlesnakes have either a dark facial bar or none. Although, admittedly, almost impossible to ascertain on a living snake, the presence of more than 2 internasal scales is a diagnostic characteristic for the prairie rattlesnake and its subspecies.

Comments: Contrary to popular belief, the prairie rattler probably does not cohabit peacefully with prairie dogs, burrowing owls, or other endotherms. Rather, these are known to be on the list of prey species.

It does, however, readily utilize unoccupied mammal or bird burrows.

Massasaugas and Pygmy Rattlesnakes; genus *Sistrurus*

Only two of the three species currently contained in this genus occur in the United States. Each species is divided into three subspecies and all may be found in eastern North America.

The members of the genus *Sistrurus* are small, often excitable rattlesnakes with slender tails and proportionately small rattles that are difficult to hear.

The origin of the size-related common name of pygmy rattler is easily fathomed but the term "massasauga" is more obscure. This name is of Chippewa derivation and means "great river mouth."

Although the drop-for-drop potency of the venom of these rattlesnakes is quite high, the fangs of the snakes are relatively short and the venom yield proportionately low.

The presence of the 9 unfragmented crown scales differentiates the pygmy rattlers and the massasaugas from the members of the typical rattlesnakes (genus *Crotalus*), which have thoroughly fragmented crown scales. The scales are keeled, the anal plate and subcaudal scales are undivided, and the pupil is vertically elliptical. The head is distinctly larger than the neck, but not overly broad.

These snakes vary in abundance. At one extreme, the eastern massasauga is becoming very uncommon. Conversely the dusky pygmy rattlesnake can be present in such immense numbers that these might even be termed concentrations.

The young of these snakes have greenish or yellowish tail tips that are used in caudal luring of prey. In the act of caudal luring, the rattlesnake coils tightly, but elevates and writhes its brightly colored tail tip next to its head. An amphibian or lizard, thinking it is going to make a meal of this "caterpillar," approaches and instead becomes a meal for the waiting snake.

Prairies, canal edges, and open but damp woodlands are among the habitats utilized by these little snakes.

Massasaugas have up to 18 neonates in a clutch, but the western pygmy rattler has a documented clutch of 32.

The rattlesnakes of this genus may vary in disposition from rather benign to completely irascible. As often as not, if approached, a coiled snake will remain quiescent and a crawling snake will freeze, both responses being tried and true ways of avoiding detection. However, if molested or injured the snakes strike readily and accurately. Some of these snakes are nervous and irritable and may begin striking while the offender is still well beyond range. When confronted, these rattlers often coil rather loosely, face their perceived adversary, and twitch their head nervously.

Although massasaugas may eat an occasional frog or lizard, rodents comprise the lion's share of the prey. Voles seem the most favored prey, with other mouse species being opportunistically accepted. Pygmy rattlers eat amphibians, lizards, an occasional smaller snake, nestling rodents, and some insects.

204. Eastern Massasauga

Sistrurus catenatus catenatus

Toxicity/Disposition: Dangerously venomous. Although bites from this rattlesnake are seldom considered life threatening, the venom is relatively potent. This snake should be left undisturbed whenever possible.

204. Eastern Massasaugas

Abundance/Range: This small rattlesnake is becoming increasingly uncommon and is listed as threatened, endangered, or a species of special concern throughout its range. It should not be harmed.

Eastern massasaugas occur in Ontario, Michigan, Wisconsin, Iowa, Illinois, Indiana, Ohio, western Pennsylvania, and western New York.

Habitat: Bogs, marshes, swamps, moist forested lowlands, and damp meadows and prairies are among the habitats utilized by these small venomous snakes. For hibernating, massasaugas seek rodent, crayfish, and stump holes, descending nearly to the level of ground water. After emerging from their hibernacula in the spring, massasaugas seek sunny, elevated spots for basking, often climbing to the tops of grass clumps, low shrubs, and if in marshy areas, beaver and muskrat lodges.

Size: Eastern massasaugas are adult at 20–26 inches. They rarely exceed 34 inches and the record size is 39½ inches. Neonates are 7–10 inches long.

Identifying features: Eastern massasaugas occur in two very different age-related color morphs. One is of earthen tones and the other is darker, the result of a variable suffusion of melanin that becomes more complete with advancing age. An old snake of this latter phase may be an almost unrelieved black.

The ground color of the lighter phase may vary from tan to medium gray. The dorsal saddles are dark brown to black and normally are narrowly outlined with light gray or white; however, the outlining is sometimes lacking. The most anterior blotches are elongate, begin immediately posterior to the large head scales, and extend onto the nape. A broad, often light-bordered dark line extends rearward from the eye to the back of the head. Dark, rounded, lateral blotches, usually also narrowly edged with light pigment, are evident. The belly is predominantly black but may have a smudged appearance.

Among neonates the melanistic examples bear the same pattern as the normally colored animals. The rattle of this species is easily seen and also easily heard, unless the snake is a newborn or has broken the rattle.

This snake has keeled scales in 25 rows; the anal plate is undivided.

Similar snakes: The timber rattlesnake is the only other species within the range of the eastern massasauga to have a rattle. The crown scales of the timber rattlesnake are fragmented. Water snakes lack the rattle. The other two races of massasauga tend to be paler and have predominantly light bellies.

Comments: Since massasaugas are still on high ground (as opposed to the low-lying swamplands where they hibernate) during the late summer haying season, they are often encountered during mowing and baling procedures. Eastern massasaugas are becoming increasingly rare throughout their range. Most states now protect them.

Additional subspecies

205. The Desert Massasauga, *Sistrurus catenatus edwardsi*, is found in a small area of southern Texas, then again from western Texas to southeastern Arizona. It is protected in the latter state. Disjunct populations occur in northern Mex-

205. Desert Massasauga

ico. It is an aridland snake, being found in grasslands and surrounding desert scrub. This is the smallest of the three races. Adults are usually 14–18 inches in length; the record size is a mere 21 inches. This slender snake, very much like the western massasauga in appearance, is of rather pale coloration and has an almost white, usually unmarked belly. Its keeled scales are in 23 rows.

206. Western Massasauga

206. The Western Massasauga, *Sistrurus catenatus tergeminus*, ranges southward from southeastern Nebraska and adjacent Iowa to eastern Texas. Although found in semi-arid to arid habitats (prairies, plains, and grasslands), this little rattlesnake is often associated with areas of surface water. The environs of bogs and prairie potholes are favored habitats. Adult at 16–24 inches, the western massasauga has an authenticated maximum length of 34¾ inches.

The ground color of this snake is tan to light brown and the 3 rows of blotches (those in the dorsal row are the largest) are dark brown. The belly is light with dark smudges. The scales are keeled and in 25 rows.

207. Carolina Pygmy Rattlesnake

Sistrurus miliarius miliarius

Toxicity/Disposition: Dangerously venomous. Although quite potent in drop-for-drop toxicity, this is a small rattler with a low venom yield. Bites sustained by an otherwise healthy adult human are not usually considered life threatening.

Abundance/Range: Although it is protected in North Carolina, the Carolina pygmy rattlesnake is not (at the time of this writing) protected by the other states in which it is found. This is a moderately common snake.

207. Carolina Pygmy Rattlesnakes

Carolina
dusky
western

    The Carolina pygmy rattlesnake ranges southward from coastal central North Carolina, through all but extreme western South Carolina, westward through central Georgia to central and northwestern Alabama.

Habitat: On the coastal plain this is considered a pinewoods/scrub oak species. In the Piedmont and elsewhere it inhabits both pine and open mixed woodlands. It occurs near water sources as well as in dryer areas. It occurs also in suburban and rural areas replete with yard litter, along the edges of dumps, and on country roadways on warm evenings. In bygone years we have found this snake in the proximity of transient sawmills. It was especially numerous where slabs of bark, trunk, and sawdust piles had been left littering an area. Today (2004), because of changed lumbering practices, the sawmills, the artificial habitat they created, and the concentrations of Carolina pygmy rattlesnakes are all things of the past.

Size: The Carolina pygmy rattlesnake is adult at 14–20 inches and has a verified record size of only 25 inches. Neonates are 5½–7½ inches long.

Identifying features: In coloration, the Carolina race is the most variable of the three subspecies of pygmy rattlers. In the mixed woodlands of central eastern

North Carolina the ground color of most specimens is pinkish, light reddish brown, or russet. The dorsal blotches are well defined and dark brown to dark reddish brown. The lateral blotches are well separated from the row of dorsal blotches. In some areas of central North Carolina the ground color is a rather distinctive grayish lavender and the dorsal and lateral blotches are very dark brown to black. In other areas the ground color is a light gray, but the blotching remains very dark brown to black. The dark blotches are usually narrowly—sometimes *very* narrowly—bordered with light pigment. In all cases, whether the snake is of the red, the light gray, or an intermediate phase, a prominent orange vertebral stripe is usually visible between the dark blotches. The vertebral stripe is often most intense anteriorly. An elongate pair of dark blotches runs from the supraocular scales to the anterior nape and a dark blotch extends from each eye to the rear of the head. The venter is usually lighter than the dorsum and bears paired, but variably shaped dark blotches.

Neonates are usually more intensely colored and patterned than the adults and have a yellowish green tail tip.

They intergrade with the dusky pygmy rattlesnake along the southern periphery of their range and with the western pygmy rattlesnake on the western periphery.

Similar species: Throughout its range, this is the only small rattlesnake with large crown scales. Water snakes and copperheads lack the rattle.

Additional subspecies

208. The Dusky Pygmy Rattlesnake, *Sistrurus miliarius barbouri*, is the southeasternmost representative of pygmy rattler. It ranges along the coastal plains from southeastern South Carolina to western Alabama and is found throughout Florida. It is common to abundant through most of its range. This is a snake of the pinewoods/scrub oak regions as well as open, mixed, woodlands. The dusky pygmy may be found in a wide spectrum of habitats. It occurs on wet prairies, near the environs of marshes, at swamp edges, on road verges that parallel drainage canals, and near agricultural areas. It often thrives in suburban and rural areas replete with yard litter and along the edges of dumps, and may be seen crossing roadways on warm evenings.

Slightly larger than the Carolina race, the dusky pygmy rattlesnake is adult at 14–20 inches in length and has a verified record size of only 31½ inches.

This is a dark yet prominently patterned little rattlesnake. The ground color of most specimens is some shade of gray. There is a row of prominent black dor-

208. Dusky Pygmy Rattlesnake

sal blotches as well as a row of lateral blotches on each side. An orange vertebral stripe is usually present (at least anteriorly) between the dark dorsal blotches. The lateral blotches are well separated from the row of dorsal blotches.

Neonates are usually more intensely colored and patterned than the adults and have a yellowish green tail tip. This may be retained well into adulthood.

209. The Western Pygmy Rattlesnake, *Sistrurus miliarius streckeri,* can vary in abundance from common to rare. It is found from eastern Louisiana and central Kentucky to eastern Texas and eastern Oklahoma. Look for it near prairie ponds, beaver lakes, marshes, and swamp edges or canals.

This, too, is a small race, topping out at a hair over 25 inches but more commonly measuring 15–20 inches long. This is a pretty form with a pale gray, grayish brown, or olive yellow ground color and well-separated, narrow dorsal blotches, and alternating rounded lateral blotches. A third row of even smaller blotches occurs just above the belly plates; the lower edges of these blotches

209. Western Pygmy Rattlesnake

reach onto the belly plates. The light area between the dorsal blotches is usually orangish. The belly is mostly light but has some scattered dark areas.

Up to 32 neonates have been reported for this tiny rattlesnake. Most clutches contain 3–15 babies.

Glossary

Aestivation A period of warm weather inactivity, often triggered by excessive heat or drought.

Ambient temperature The temperature of the surrounding environment.

Anterior Toward the front.

Anus The external opening of the cloaca; the vent.

Apical pits Tiny depressions on the dorsal and lateral scales of a snake.

Arboreal Of or relating to trees as habitat; dwelling in trees.

Brille The clear spectacle that protects the eyes of a snake.

Caudal Pertaining to the tail.

Caution colors Red = stop; yellow = caution.

Cloaca The common chamber into which digestive, urinary and reproductive systems empty and which itself opens exteriorly through the vent or anus.

Clutch A group of babies or eggs originating from a single deposition. Used here for egg-laying species.

Congeneric Grouped in the same genus.

Crepuscular Active at dusk and/or dawn.

Deposition As used here, the laying of the eggs or birthing of young.

Deposition site The nesting or birthing site.

Dichromatic Exhibiting two color phases, often sex-linked.

Dimorphic A difference in form, build, or coloration involving the same species, often sex-linked.

Diurnal Active in the daytime.

Dorsal Pertaining to the back, the upper surface.

Dorsolateral Pertaining to the upper side.

Dorsum The upper surface.

Ecdysis Shedding of skin.

Ecological niche A species' "overall place" in an ecosystem.

Ectothermic "Cold-blooded," pertaining to an organism that absorbs heat from the environment.

Endemic Confined to a specific region.

Endothermic "Warm-blooded," pertaining to an organism that produces its own body heat.

Exotic Nonnative, introduced from other geographic locations.

Form An identifiable species or subspecies.

Fossorial Burrowing.

Genus (pl. **Genera**) A taxonomic classification of a group of species having similar characteristics. The genus falls between the next higher designation of Family and the next lower designation of Species. The first letter of the generic name is always capitalized and the name is always written in italics.

Gravid The reptilian equivalent of mammalian pregnancy.

Ground color The predominating body color.

Gular Pertaining to the throat.

Hatchling A snake newly emerged from the egg.

Heliothermic Pertaining to a species that basks in the sun to thermoregulate.

Herpetology The study (often scientifically oriented) of reptiles and amphibians.

Hybrid Offspring resulting from the breeding of two species or noncontiguous subspecies.

Intergrade Breeding of two adjacent subspecies, and the resultant offspring.

Interstitial Skin between the scales.

Juvenile A young or immature specimen.

Keel A longitudinal ridge (or ridges) on the scales.

Lateral Pertaining to the side.

Litter A group of babies originating from a single deposition.

Mandibles Jaws.

Mandibular Pertaining to the jaws.

Melanism A profusion of black pigment.

Microhabitat The precise habitat utilized by a species.

Middorsal Pertaining to the middle of the back.

Midventral Pertaining to the center of the belly.

Monotypic Containing but one type.

Neonate A newborn snake; usually used only in reference to live-bearing species.

Nocturnal Active at night.

Nominate The first named form.

Nuchal Pertaining to the nape.

Ocelli Dark or light-edged circular spots.

Ontogenetic Age-related (color) changes.

Ophiophagous Feeding on snakes.

Oviparous Reproducing by means of eggs that hatch after laying.

Paravertebral Pertaining to the area on each side of the vertebral column.

Parthenogenesis Reproduction without fertilization.

Parturition Birthing.

Photoperiod The daily/seasonally variable length of the hours of daylight.

Poikilothermic (also **ectothermic**) Having no internal body temperature regulation; another old term for "cold-blooded."

Postocular Behind the eye.

Race Among snakes, a subspecies.

Rostral scale The enlarged scale of the tip of the snout.

Scute A large scale or plate; among snakes, primarily the wide ventral scales.

Sibling species Two or more similar appearing species supposedly derived from the same parental stock. Sibling species are often unidentifiable in the field.

Species (abbr. **sp.**) A group of similar organisms that produce viable young when breeding. The taxonomic designation that falls beneath Genus and above Subspecies.

Subcaudal The underside of the tail.

Subocular Below the eye.

Subspecies (abbr. **ssp.**) The subdivision of a species. A race that may differ slightly in color, size, scalation, or other criteria.

Subsurface Beneath the surface.

Supralabials The upper lip scales.

Supraocular Above the eye.

Sympatric Occurring together.

Taxonomy The science of classification of plants and animals.

Terrestrial Land-dwelling.

Thermoregulate To regulate (body) temperature; in an ectotherm, by choosing a warmer or cooler environment.

Vent The external opening of the cloaca; the anus.

Venter The underside; the belly.

Ventral Pertaining to the undersurface or belly.

Ventrolateral Pertaining to the sides of the belly.

Vertebral Pertaining to the center of the back; mid dorsal.

Viviparous Bearing live young.

References and Additional Reading

A few of the pertinent titles pertaining to the snakes of the eastern and central United States and Canada:

Ashton, R. E., Jr., S. R. Edwards, and G. R. Pisani. 1976. *Endangered and Threatened Amphibians and Reptiles of the United States.* Society for the Study of Amphibians and Reptiles, Circular 5. Lawrence, Kans.

Ashton, R. E., Jr., and P. S Ashton. 1981. *Handbook of Reptiles and Amphibians of Florida.* Part I, *The Snakes.* Miami: Windward Publishing.

Bartlett, R. D. 1988. *In Search of Reptiles and Amphibians.* New York: E. J. Brill.

Bartlett, R. D., and Patricia P. Bartlett. 1998. *Snakes: A Complete Pet Owner's Manual.* Hauppauge, N.Y.: Barron's Educational Series.

————. 1999. *Terrarium and Cage Construction and Care.* Hauppauge, N.Y.: Barron's Educational Series.

Bartlett, R. D., and Ronald G. Markel. 1995. *Kingsnakes and Milksnakes: A Complete Pet Owner's Manual.* Hauppauge, N.Y.: Barron's Educational Series.

Behler, John L., and F. Wayne King. 1979. *The Audubon Society Field Guide to North American Reptiles and Amphibians.* New York: Alfred Knopf.

Blaney, R. M. 1977. Systematics of the common kingsnake, *Lampropeltis getulus. Tulane Zoological Studies 19* (nos. 3–4):47–103.

Boundy, J. 1997. *Snakes of Louisiana.* Baton Rouge: Louisiana Deptartment of Wildlife and Fisheries.

————. 1995. Maximum lengths of North American snakes. *Bulletin of the Chicago Herpetological Society* 30:109–122.

Brown, C. W., and C. H. Ernst. 1986. A study of variation in eastern timber rattlesnakes, *Crotalus horridus* Linnae (Serpentes: Viperidae). *Brimleyana* 12:57–74.

Burbrink, F. T. 2002. Phylogenetic analysis of the corn snake (*Elaphe guttata*) complex as inferred from maximum likelihood and Bayesian analyses. *Molecular Phylogenetics and Evolution* 25:465–476.

————. 2001. Systematics of the eastern ratsnake complex. *Herpetological Monographs* 15:1–53.

Collins, Joseph T. 1991. Viewpoint: A new taxonomic arrangement for some North American amphibians and reptiles. *Herpetological Review* 22:42–43.

————. 1982. *Amphibians and Reptiles in Kansas.* 2nd ed. Lawrence: University of Kansas.

Conant, Roger, and Joseph T. Collins. 1991. *A Field Guide to The Reptiles and Amphibians of Eastern and Central North America.* 3rd ed. Boston: Houghton Mifflin.

Crother, Brian I. (Chair). 2000. *Scientific and Standard English Names of Amphibians and Reptiles of North America north of Mexico, with Comments Regarding Confidence in our Understanding.* Society for the Study of Amphibians and Reptiles, Circular 29. Lawrence, Kans.

Degenhardt, William G., Charles W. Painter, and Andrew H. Price. 1996. *Amphibians and Reptiles of New Mexico.* Albuquerque: University of New Mexico Press.

DeGraaf, Richard M., and Deborah D. Rudis. 1983. *Amphibians and Reptiles of New England.* Amherst: University of Massachusetts Press.

Dixon, James R. 1987. *Amphibians and Reptiles of Texas.* College Station: Texas A&M University Press.

Dundee, Harold A., and Douglas A. Rossman. 1989. *The Amphibians and Reptiles of Louisiana.* Baton Rouge: Louisiana State University Press.

Ernst, Carl H. 1992. *Venomous Reptiles of North America.* Washington, D.C.: Smithsonian Institution Press.

Ernst, Carl H., and Evelyn M. Ernst. 2003. *Snakes of the United States and Canada.* Washington, D.C.: Smithsonian Institution Press.

Gloyd, Howard K. 1978. *The Rattlesnakes, Genera* Sistrurus *and* Crotalus. Lawrence, Kans.: Society for the Study of Amphibians and Reptiles.

Gloyd, Howard K., and Roger Conant. 1990. *Snakes of the Agkistrodon Complex, A Monographic Review.* Ithaca, N.Y.: Society for the Study of Amphibians and Reptiles.

Green, N. Bayard, and Thomas K. Pauley. 1987. *Amphibians and Reptiles in West Virginia.* Pittsburgh: University of Pittsburgh Press.

Greene, Harry W. 1997. *Snakes: The Evolution of Mystery in Nature.* Berkeley and Los Angeles: University of California Press.

Halliday, Tim, and Kraig Adler (eds.). 1986. *The Encyclopedia of Reptiles and Amphibians.* New York: Facts on File.

Harding, James H. 1997. *Amphibians and Reptiles of the Great Lakes Region.* Ann Arbor: University of Michigan Press.

Hunter, Malcolm L., Aram J. K. Calhoun, and Mark McCollough (eds.) 1999. *Maine Amphibians and Reptiles.* Orono: University of Maine Press.

Johnson, Tom R. 1987. *The Amphibians and Reptiles of Missouri.* Jefferson City: Missouri Department of Conservation.

Klauber, Lawrence M. 1972. *Rattlesnakes: Their Habits, Life Histories, and Influence on Mankind.* 2nd ed. Berkeley and Los Angeles: University of California Press.

Lazell, James D., Jr. 1976. *This Broken Archipelago: Cape Cod and the Islands, Amphibians and Reptiles.* New York: Quadrangle/New York Times Book Co.

Levell, John P. 1995. *A Field Guide to Reptiles and the Law.* Excelsior, Minn.: Serpent's Tale.

Martof, Bernard S., William M. Palmer, Joseph R. Bailey, and Julian R. Harrison III. 1980. *Amphibians and Reptiles of the Carolinas and Virginia.* Chapel Hill: University of North Carolina Press.

McDarmid, Roy W., Jonathan A. Campbell, and T'Shaka A. Toure. 1999. *Snake Species of the World: A Taxonomic and Geographic Reference.* Vol. 1. Washington, D.C.: Herpetologists' League.

Mehrtens, John M. 1987. *Living Snakes of the World in Color.* New York: Sterling.

Miller, D. J. 1979. *A Life History Study of the Gray-banded Kingsnake,* Lampropeltis mexicana alterna, *in Texas.* Alpine, Tex.: Chihuahuan Desert Research Institute.

Minton, Sherman A., Jr. 2001. *Amphibians and Reptiles of Indiana.* Indianapolis: Indiana Academy of Science.

Mitchell, Joseph C. 1994. *The Reptiles of Virginia.* Washington, D.C.: Smithsonian Institution Press.

Moler, Paul E. (ed.). 1992. *Rare and Endangered Biota of Florida.* Vol. 3, *Amphibians and Reptiles.* Gainesville: University Press of Florida.

—————. 1990. *A Checklist of Florida's Amphibians and Reptiles.* Rev. ed. Tallahassee: Florida Game and Fresh Water Fish Commission.

Mount, Robert H. 1975. *The Reptiles and Amphibians of Alabama.* Auburn: Auburn University Agricultural Experiment Station.

Oldfield, Barney, and John J. Moriarty. 1994. *Amphibians and Reptiles Native to Minnesota.* Minneapolis: University of Minnesota Press.

Palmer, William M., and Alvin L. Braswell. 1995. *Reptiles of North Carolina.* Chapel Hill: University of North Carolina Press.

Pisani, G. R., J. T. Collins, and S. R. Edwards. 1973. A re-evaluation of the subspecies of *Crotalus horridus. Transactions of the Kansas Academy of Science* 75:255–263

Roze, Janis A. 1996. *Coral Snakes of the Americas.* Malabar, Fla.: Krieger.

Rubio, Manny. 1998. *Rattlesnake: A Portrait of a Predator.* Washington, D.C.: Smithsonian Institution Press.

Schwartz, Vicki, and David M. Golden. 2002. *Field Guide to Reptiles and Amphibians of New Jersey.* Vineland: New Jersey Divison of Fish and Wildlife.

Smith, H. M., D. Chiszar, J. R. Staley II, and K. Tepedelen. 1994. Populational relationships in the corn snake *Elaphe guttata* (Reptilia: Serpentes). *Texas Journal of Science* 46:259–292

Tyning, Thomas F. 1990. *A Guide to Amphibians and Reptiles.* Boston: Little, Brown and Co.

Vogt, Richard Carl. 1981. *Natural History of Amphibuians and Reptiles of Wisconsin.* Milwaukee: Milwaukee Public Museum.

Werler, John E., and James R. Dixon. 2000. *Texas Snakes: Identification, Distribution, and Natural History.* Austin: University of Texas Press.

Wilson, Larry David, and Louis Porras. 1983. *The Ecological Impact of Man on the South Florida Herpetofauna.* Lawrence: University of Kansas.

Wright, Albert H., and Anna A. Wright. 1957. *Handbook of Snakes of the United States and Canada.* 2 vols. Ithaca, N.Y.: Comstock.

Zug, George R., Laurie J. Vitt, and Janalee P. Caldwell. 2001. *Herpetology: An Introductory Biology of Amphibians and Reptiles.* San Diego: Academic.

Index

Acrochordidae 17–19
Acrochordus javanicus (introduced species)
    18–19
Agkistrodon 287–99
    contortrix contortrix 287–90
    contortrix laticinctus 290–91
    contortrix mokasen 291–92
    contortrix phaeogaster 292–93
    contortrix pictigaster 293–94
    piscivorus piscivorus 294–99
    piscivorus conanti 297–99
    piscivorus leucostoma 299
Arizona 96–99
    elegans arenicola 99
    elegans elegans 96–98
    elegans philipi 99

Boa constrictor imperator (introduced species)
    22–23
Boa constrictor
    Colombian (introduced species) 22–23
Bogertophis subocularis 104–6
Boidae 20–27
Bullsnake 150–51, 154–55

Captivity, Snakes in 4–6
Cautionary notes 3
Carphophis 253–57
    amoenus amoenus 253–55
    amoenus helenae 255–56
    amoenus vermis 256–57
Cemophora 100–103
    coccinea coccinea 100–102
    coccinea copei 102
    coccinea lineri 103
Clonophis kirtlandi 164–66

Coachwhip 39–42
    Eastern 39–40
    Lined 41
    Western 41–42
Coluber 28–38
    constrictor anthicus 32–33
    constrictor constrictor 31–32
    constrictor etheridgei 33
    constrictor flaviventris 34
    constrictor foxii 35
    constrictor helvigularis 35
    constrictor latrunculus 36
    constrictor oaxaca 36
    constrictor paludicola 37
    constrictor priapus 37–38
Colubrinae 28–84
Coniophanes imperialis imperialis 85–87
Copperhead 287–94
    Broad-banded 290–91
    Northern 291–92
    Osage 292–93
    Southern 277–90
    Trans-Pecos 293–94
Cottonmouth 287, 294–99
    Eastern 294–97
    Florida 297–99
    Western 299
Crotalinae 285–327
Crotalus 300–317
    adamanteus 302–3
    atrox 303–6
    horridus atricaudatus 308–10
    horridus horridus 306–8
    lepidus klauberi 312
    lepidus lepidus 310–11
    molossus molossus 312–14

Crotalus—continued
  scutulatus scutulatus 314–15
  viridis viridis 316–17

Diadophis 257–63
  punctatus acricus 259–60
  punctatus arnyi 260–61
  punctatus edwardsi 261–62
  punctatus punctatus 258–59
  punctatus regalis 262–63
  punctatus stictogenys 263
Dipsadinae 84–94
Drymarchon 47–50
  corais couperi 48–49
  corais erebennus 50
Drymobius margaritiferus margaritiferus 50–52

Elaphe 107–24
  bairdi 108–9
  guttata emoryi 114
  guttata guttata 109–13
  guttata meahllmorum 114
  obsoleta lindhemeri 117–18
  obsoleta obsoleta 115–17
  obsoleta quadrivittata x *o. spiloides* 121
  obsoleta quadrivittata 118–19
  obsoleta rossalleni 119–20
  obsoleta spiloides 120–21
  vulpina gloydi 123–24
  vulpina vulpina 122–23
Elapidae 279–85

Farancia 264–69
  abacura abacura 265–66
  abacura reinwardtii 266–67
  erytrogramma erytrogramma 267–68
  erytrogramma seminola 269
Ficimia streckeri 62–63

Gyalopion canum 63–65

Heterodon 270–78
  nasicus gloydi 273
  nasicus kennerlyi 274
  nasicus nasicus 270–73
  platirhinos 274–76
  simus 276–78
Hypsiglena torquata jani 87–90

Kingsnake 124–50
  Apalachicola Lowland 137–39
  California (introduced species) 134–35
  Desert 140–41
  Eastern 132–34
  Eastern Black 140
  Florida 135–36
  Gray-banded 126–28
  Mole 131–32
    South Florida 130–31
  Outer Banks 141–42
  Peninsula 136–37
  Prairie 129–30
  Scarlet 146–47
  Speckled 139

Lampropeltinae 94–163
Lampropeltis 124–51
  alterna 126–28
  calligaster calligaster 129–30
  calligaster occipitolineata 130–31
  calligaster rhombomaculata 131–32
  getula californiae (introduced species)
    134–35
  getula floridana 135–36
  getula floridana x *L. g. getula* 136–37
  getula getula 132–34
  getula, unresolved affinities 137–39
  getula holbrooki 139
  getula nigra 140
  getula splendida 140–41
  getula sticticeps 141–42
  triangulum amaura 144–45
  triangulum annulata 145
  triangulum celaenops 145–46
  triangulum elapsoides 146–47
  triangulum gentilis 147
  triangulum multistriata 148
  triangulum syspila 148–49
  triangulum ssp., unresolved affinities
    149–50
  triangulum triangulum 142–44
Leptodeira septentrionalis septentrionalis 90–92
Leptotyphlopidae 10–14
Leptotyphlops 10–14
  dulcis dulcis 12–13
  dulcis dissectus 13
  humilis segregus 14

Massasauga 319–22
    Desert 321–22
    Eastern 319–21
    Western 322
Masticophis 38–47
    flagellum flagellum 39–40
    flagellum lineatulus 41
    flagellum testaceus 41–42
    schotti ruthveni 44–45
    schotti schotti 42–44
    taeniatus girardi 46
    taeniatus taeniatus 45–46
Micrurus 282–85
    fulvius 282–84
    tener 284–85

Natricinae 163–252
Nerodia 167–93
    clarkii clarkii 168–69
    clarkii compressicauda 169–70
    clarkii taeniata 170–71
    cyclopion 171–73
    erythrogaster erythrogaster 173–75
    erythrogaster flavigaster 175–76
    erythrogaster neglecta 176–77
    erythrogaster transversa 177
    fasciata confluens 179–80
    fasciata fasciata 178–79
    fasciata pictiventris 180–81
    floridana 181–82
    harteri harteri 182–84
    harteri paucimaculata 184–85
    rhombifer rhombifer 185–86
    sipedon insularum 189
    sipedon pleuralis 189–90
    sipedon sipedon 187–88
    sipedon williamengelsi 190–91
    taxispilota 191–93

Opheodrys 52–56
    aestivus aestivus 53–54
    aestivus carinatus 54–55
    vernalis 55–56

Pituophis 150–59
    catenifer affinis 151–53
    catenifer ruthveni 153–54
    catenifer sayi 154–55

melanoleucus lodingi 157–58
melanoleucus melanoleucus 155–57
melanoleucus mugitus 158–59
Pit Viper 285–327
Python 23–27
    Ball (introduced species) 25–27
    Burmese (introduced species) 24–25
Python 23–27
    molurus bivittatus (introduced species) 24–25
    regius (introduced species) 25–27

Racer 29–38
    Black
        Northern 31–32
        Southern 37–38
    Black-masked 36
    Blue 35
    Brown-chinned 35
    Buttermilk 32–33
    Eastern Yellow-bellied 34
    Everglades 37
    Mexican 36
    Northern Speckled 51–52
    Tan 33
Ramphotyphlops braminus (introduced species) 15–16
Rattlesnake 300–327
    Canebrake 308–10
    Diamond-backed 302–6
        Eastern 302–3
        Western 303–6
    Mohave 314–15
    Northern Black-tailed 312–14
    Prairie 316–17
    Pygmy 322–27
        Carolina 322–25
        Dusky 325–26
        Western 326–27
    Rock 310–12
        Banded 312
        Mottled 310–11
    Timber 306–8
Regina 193–202
    alleni 194–95
    grahamii 195–97
    rigida deltae 199
    rigida rigida 197–99

Regina—continued
  rigida sinicola 199–200
  septemvittata 200–202
Rhadinea flavilata 92–94
Rhinocheilus lecontei tessellatus 159–61

Salvadora 57–61
  grahamiae grahamiae 58–59
  grahamiae lineate 59–60
  hexalepis deserticola 60–61
Seminatrix
  pygaea cyclas 204–5
  pygaea paludis 205
  pygaea pygaea 202–4
Sistrurus 318–27
  catenatus catenatus 319–21
  catenatus edwardsi 321–22
  catenatus tergeminus 322
  miliarius barbouri 325–26
  miliarius miliarius 322–25
  miliarius streckeri 326–27
Snake
  Black-headed
    Plains 76–78
    Mexican 69–71
    Southwestern 75–76
    Trans-Pecos 72–73
  Black-striped 84–88
    Tamaulipan 86–88
  Blind 10–16
    Brahminy (introduced species) 15–16
    New Mexican 13
    Slender 10–14
    Texas 12–13
    Trans-Pecos 14–15
    Typical 15–16
  Brown 206–11
    Florida 210
    Marsh 209
    Midland 211
    Northern 206–9
    Texas 210
  Cat-eyed 90–92
    Northern 91–92
  Coral 279–85
    Eastern 282–84
    Texas 284–85
  Corn 109–13
  Crayfish 193–202

Delta 199
Glossy 197–99
Graham's 195–96
Gulf 199–200
Striped 194–95
Crowned
  Central Florida 80–81
  Coastal Dunes 82
  Peninsula 79–80
  Rim Rock 78–79
  Southeastern 71–72
Earth 247–52
  Eastern 249–50
  Mountain 251–52
  Rough 247–49
  Smooth (*see* Eastern, Mountain, or Western)
  Western 251
Elephant-trunk (*see* Snake, Javan File)
File 17
Flat-headed 74–75
Fox
  Eastern 123–24
  Western 122–24
Garter 216–45
  Black-necked 220–23
    Eastern 222–23
    Western 220–22
  Blue-striped 244–45
  Butler's 219–20
  Checkered 225–27
  Chicago 243–44
  Eastern 237–39
  Maritime 241–42
  New Mexico 240–41
  Plains 231–33
  Red-sided 242–43
  Short-headed 217–18
  Texas 240
  Wandering 223–25
Glossy 96–99
  Kansas 96–98
  Painted Desert 99
  Texas 99
Gopher 150–51
  Sonoran 151–53
Green 52–56
  Rough
    Florida 54–55
    Northern 53–54

Smooth 55–56
Ground
  Great Plains 64–68
  Taylor's 68
Hog-nosed 252–53, 270–78
  Dusty 273
  Eastern 274–76
  Mexican 274
  Plains 270–73
  Southern 276–78
Hook-nosed 61–64
  Chihuahuan 63–64
  Tamaulipan 62–63
Indigo 47–50
  Eastern 48–49
  Texas 50
Javan File (introduced species) 18–19
King (see Kingsnake)
Kirtland's 165–66
Lined 245–47
Long-nosed
  Texas 159–61
Lyre
  Texas 82–84
Milk 142–51
  Central Plains 147
  Coastal Plains 149–50
  Eastern 142–44
  Louisiana 144–45
  Mexican 145
  New Mexican 145–46
  Pale 148
  Red 148–49
Mud 264–67
  Eastern 265–66
  Western 266–67
Night 87–88
  Texas 88–90
Patch-nosed 57–61
  Big Bend 60–61
  Mountain 58–59
  Texas 59–60
Pine 150–51
  Black 157–58
  Florida 158–59
  Louisiana 153–54
  Northern 155–57
Pine Woods 92–94
Queen 200–202

Rainbow 264–65, 267–69
  Common 267–68
  South Florida 269
Rat 103–24
  Baird's 108–9
  Black 115–17
  Deckert's (see Rat, Yellow)
  Everglades 119–20
  Gray 120–21
  Great Plains 114
  Gulf Hammock 121
  Southwestern 114–15
  Texas 117–18
  Trans-Pecos 104–6
  Yellow 118–19
Red-bellied 212–16
  Black Hills 215–16
  Florida 214–15
  Northern 212–14
Ribbon
  Aridland 229–30
  Blue-striped 235
  Eastern 233–35
  Gulf Coast 230
  Northern 236–37
  Orange-striped 227–29
  Peninsula 236
  Red-striped 231
  Western (see Ribbon, Orange-striped)
Ring-necked 252–53, 257–65
  Keys 259–60
  Mississippi 263–64
  Northern 261–62
  Prairie 260–61
  Regal 262–63
  Southern 258–59
Salt Marsh
  Atlantic 170–71
  Gulf 168–69
  Mangrove 169–70
Scarlet 100–103
  Florida 100–102
  Northern 102
  Texas 103
Short-tailed 161–63
Swamp 202–5
  Carolina 205
  North Florida 202–4
  South Florida 204–5

Water 167–93
  Banded 178–79
  Blotched 177
  Brazos 182–84
  Broad-banded 179–80
  Brown 191–93
  Carolina 190–91
  Concho 184–85
  Copper-bellied 176–77
  Diamond-backed 185–86
  Florida 180–81
  Green
    Florida 181–82
    Mississippi 171–73
  Lake Erie 189
  Midland 189–90
  Northern 187–88
  Red-bellied 173–75
  Yellow-bellied 175–76
  Worm 252–57
    Eastern 253–55
    Midwest 255–56
    Western 256–57
Sonora 65–68
  semiannulata semiannulata 65–67
  semiannulata taylori 68
Stilosoma extenuatum 161–63
Storeria 205–16
  dekayi dekayi 206–9
  dekayi limnetes 209
  dekayi texana 209–10
  dekayi victa 210–11
  dekayi wrightorum 211
  occipitomaculata obscura 214–15
  occipitomaculata occipitomaculata 212–14
  occipitomaculata pahasapae 215–16

Tantilla 68–81
  atriceps 69–70
  coronata 71–72
  cucullata 72–73
  gracilis 74–75
  hobartsmithi 75–76
  nigriceps 76–78
  oolitica 78–79

  relicta neilli 80–81
  relicta pamlica 81
  relicta relicta 79–80
Thamnophis 216–45
  brachystoma 217–18
  butleri 219–20
  cyrtopsis cyrtopsis 220–22
  cyrtopsis ocellatus 222–23
  elegans vagrans 223–25
  marcianus marcianus 225–27
  proximus diabolicus 229–30
  proximus orarius 230–31
  proximus proximus 227–29
  proximus rubrilineatus 231
  radix 231–33
  sauritus nitae 235
  sauritus sackenii 236
  sauritus sauritus 233–35
  sauritus septentrionalis 236–37
  sirtalis annectens 240
  sirtalis dorsalis 240–41
  sirtalis pallidulus 241–42
  sirtalis parietalis 242–43
  sirtalis semifasciatus 243–44
  sirtalis similes 244–45
  sirtalis sirtalis 237–39
Trimorphodon biscutatus vilkinsonii 82–84
Tropidoclonion lineatum 245–47
Typhlopidae 15–16

Viperidae 285–327
Vipers 285–327
Virginia 247–52
  striatula 247–49
  valeriae elegans 250
  valeriae pulchra 251–52
  valeriae valeriae 249–50

Whipsnakes 38–46
  Central Texas 46
  Desert 45–46
  Ruthven's 44–45
  Schott's 42–44

Xenodontinae 252–78

Richard D. and Patricia Bartlett have coauthored numerous books, including *A Field Guide to Florida Reptiles, Reptiles and Amphibians of the Amazon,* and *Florida Snakes.* Together they lead interactive tours to many areas of the Amazon Basin. Richard has published more than 500 articles about herpetology in *Tropical Fish Hobbyist, Reptiles, Reptile and Amphibian,* and others. Patricia is former director of the Fort Myers Historical Museum.

Related-interest titles from University Press of Florida

30 EcoTrips in Florida: The Best Nature Excursions (and How to Leave Only Your Footprints)
Holly Ambrose

A Field Guide and Identification Manual for Florida and Eastern U.S. Tiger Beetles
Paul M. Choate, Jr.

Florida's Snakes: A Guide to Their Identification and Habitats
R. D. Bartlett and Patricia Bartlett

Guide and Reference to the Amphibians of Eastern and Central North America, North of Mexico
R. D. Bartlett and Patricia Bartlett

Guide and Reference to the Crocodilians, Turtles, and Lizards of Eastern and Central North America, North of Mexico
R. D. Bartlett and Patricia Bartlett

A Guide to the Birds of the Southeastern States: Florida, Georgia, Alabama, and Mississippi
John H. Rappole

Hiker's Guide to the Sunshine State
Sandra Friend

Reptiles and Amphibians of the Amazon: An Ecotourist's Guide
R. D. Bartlett and Patricia Bartlett

Wild Orchids of the Canadian Maritimes and Northern Great Lakes Region
Paul Martin Brown with drawings by Stan Folsom

Wild Orchids of North America, North of Mexico
Paul Martin Brown with drawings by Stan Folsom

Wild Orchids of the Southeastern United States, North of Peninsular Florida
Paul Martin Brown with drawings by Stan Folsom

The Windward Road: Adventures of a Naturalist on Remote Caribbean Shores
Archie Carr

For more information on these and other books, visit our website at www.upf.com.